Health Research:
The Systems Approach

CONTRIBUTORS

June C. Abbey

Lawrence E. Berg

Ludwig von Bertalanffy

Ann J. Buckeridge

M. Jean Daubenmire

Kenyon B. De Greene

Gordon Hearn

Joseph W. Hess

Daniel Howland

Richard C. Jelinek

Imogene M. King

George J. Klir

Maura Kolins

Robert D. LaRue

Mihajlo D. Mesarovic

Ruben Meyer

Joyce Y. Passos

Lillian M. Pierce

Nicholas D. Rizzo

Hjalmar Rosen

Steven H. Schwartz

Grayce M. Sills

Roger I. Simon

Shirley A. Smoyak

James B. Tenney

Thomas Weisman

A. Dawn Zagornik

Health Research: The Systems Approach

EDITED BY

Harriet H. Werley, Ph.D.
Associate Dean for Research, College of Nursing
University of Illinois at the Medical Center, Chicago

Ann Zuzich, M.S.N.
Associate Professor, College of Nursing
Wayne State University, Detroit, Michigan

Myron Zajkowski, Ph.D.
Associate Professor, Department of Industrial Engineering
State University of New York at Buffalo

A. Dawn Zagornik, M.S.N.
Associate Professor, College of Nursing
Wayne State University, Detroit, Michigan

SPRINGER PUBLISHING COMPANY
New York

Springer Publishing Company, Inc.

200 Park Avenue South
New York, N.Y. 10003

76 77 78 79 80 / 10 9 8 7 6 5 4 3 2 1

Library of Congress Cataloging in Publication Data

Main entry under title:

Health research, the systems approach.

   Papers presented at a conference conducted in March 1971 by
the Center for Health Research to commemorate the 25th anniversary
of the Wayne State University College of Nursing.
   Includes index.
   1.   Public   health   research — Congresses.   2.   System
analysis — Congresses.   3.   Medical   care — Congresses.   4.   Nursing
research — Congresses.   I.   Werley,   Harriet,   H.   II.   Wayne   State
University, Detroit. Center for Health Research. III. Wayne State
University, Detroit. College of Nursing.

RA440.6.H43   362.1'01'8   73-92207
ISBN 0-8261-1710-4

Printed in the United States of America

Contents

Preface

The papers included in this volume were presented at an interdisciplinary conference on Health Research and the Systems Approach, which was conducted in March 1971 by the Center for Health Research to commemorate the twenty-fifth anniversary of the Wayne State University College of Nursing.

The major purpose of the conference was to provide a medium for information exchange between health professionals and systems specialists interested in studying health delivery systems that employ the general systems approach. Initially, we planned to invite individuals who were using the systems approach in their research or planning to do so. Accordingly, invitations were extended to systems specialists, nurses, physicians, social workers, behavioral scientists, and health administrators. However, we also received many requests for invitations from educators involved in curriculum revision in which the systems approach was to be employed. Consequently, the number of conferees actually involved in health research employing the systems approach was relatively small; to some extent this changed the tenor of the conference toward more discussion of systems concepts than of research.

The conference was designed to provide the conferees with a logical and systematic coverage of systems concepts and research in health care. Our intention in opening the conference with a presentation of general systems theory was to provide a common foundation for subsequent discussion. Ludwig von Bertalanffy was to have given the initial paper himself, but he suffered a heart attack the week before the conference and could not be present.

He asked his colleague Nicholas D. Rizzo to substitute for him. This initial paper was followed by a series of presentations dealing with selected major subsystems of health care or services. Among the subsystems discussed were the client as the focal subsystem, the professional intervention subsystems, and the administrative and support subsystems. The conference then focused on current research activities involving a group using an adaptive health systems model and other investigators working in specialized areas of interest. Because it was felt that the foregoing sessions would raise a number of methodological and measurement questions related to systems research, they were followed by a session that was devoted to discussion of the areas of systems analysis, systems synthesis, and systems evaluation. The final session of the conference was devoted to an examination of research possibilities which emerged during the presentation of papers and ensuing discussions. The papers appear in the order given at the conference with brief editorial introductions between the several sections of the papers. In the final paper the editors have attempted to summarize some of the research directions that seemed to emerge either as papers were presented or as discussions ensued.

The editors are indebted to the authors of the papers and to the other invited conferees for their contributions during the general and small group meetings.

Many persons helped to make the conference possible. The committee that formulated the plans for the conference included Ann Zuzich, A. Dawn Zagornik, and Harriet H. Werley from the College of Nursing, and Myron M. Zajkowski from the department of psychology. Dr. Werley, director of the Center for Health Research, served as conference director. Frank Misplon from the McGregor Memorial Conference Center was very helpful with arrangements that insured smooth conduct of the conference. Linda Grant provided editorial service in preparing the manuscript for presentation to the press. Gertrude Buelow and Carolyn Broughton, administrative assistant and secretary respectively from the Center for Health Research, assumed responsibility for many of the details related to planning the conference, making arrangements for program participants, assisting with the conduct of the conference, verifying references, and preparing the material for this volume. Through her assistance in all aspects of the Center's endeavors, Fredericka Shea, assistant director of the Center for Health Research, provided invaluable direct and indirect support required to conduct the conference and bring about publication of the conference papers. Charlotte Boulton, administrative secretary at the College of Nursing, University of Illinois at the Medical Center, Chicago, assisted Dr. Werley (now relocated) with final checking and other details involved in working with the publisher to bring the volume to fruition.

Financial support for the conference was obtained from Wayne State

University by Margaret L. Shetland, dean of the College of Nursing. Since the Center for Health Research was partially supported by Grant NU 361 from the Division of Nursing, National Institutes of Health, HEW, funds from this source provided for conference planning, cost of manuscript preparation, and a portion of the publication costs. In addition, some of the volume preparation expenses were partially defrayed by Grant RR 5718 from the Division of Research Resources of NIH.

Contributors

June C. Abbey is assistant professor and acting chairperson of the department of nursing in biological dysfunction at the University of California, having received her B.S. and M.S. degrees from the same institution. She expects to receive her Ph.D. degree in the spring of 1976. Professor Abbey has published many articles in the field of nursing.

Lawrence E. Berg is a consultant in planning and systems analysis. He holds a B.S. degree in economic analysis from Ohio State University. Mr. Berg is chairperson of the systems and procedure seminars conducted by the American Management Association and is deputy director of the Health Program Systems Center of the Indian Health Service at Tuscon, Arizona, where he was formerly chief of the office of systems and development. He has assisted many major cities in developing community-wide health programs.

Ludwig von Bertalanffy was a professor on the Faculties of Natural and Social Sciences at the State University of New York at Buffalo at the time of his death on June 12, 1972. He had taught biology at virtually every level, but mostly at the advanced or graduate level, and, in his last position, was a member of the teaching staff of both the departments of psychology and zoology at Buffalo. Dr. von Bertalanffy was particularly interested in interdisciplinary integration and education. His contributions to behavioral science and to theoretical psychology and psychopathology have found application in psychology, psychiatry, and related fields. He was the author of some 270 scientific publications and the editor of *General Systems: Yearbook of the Society for General Systems Research* from 1956 until his death.

Dr. von Bertalanffy's experimental work included investigations in the fields of comparative physiology of growth and metabolism, cell physiology, experimental embryology, cytochemistry, cancer research, and related fields. Among the results of his research are:

the "Bertalanffy equations" of animal growth, a concept used in fisheries biology throughout the world

the concept of an organism as an "open system," which has led to important developments in kinetics, thermodynamics, and biophysics

the foundation of "general systems theory" as an interdisciplinary field for use particularly in biological and behavioral realms to which conventional physics is inapplicable

fluorescence cytodiagnosis for early detection of cancer which, because of its many advantages, is widely used by physicians, hospitals, clinics, and cancer centers

The distinguished Dr. von Bertalanffy received his Ph.D from the University of Vienna. He was an honorary Fellow of the American Psychiatric Association, a Fellow of the American Association for the Advancement of Science and of the International Academy of Cytology, and the recipient of numerous research grants. In addition, Dr. von Bertalanffy was a founder and vice-president (1956-1960) of the Society for General Systems Research, and he held memberships in the Deutsche Akademie der Naturforscher, the New York Academy of Sciences, the American Society of Naturalists, and the Canadian Physiological Society.

Ann J. Buckeridge is associate professor of nursing at Ohio State University. Professor Buckeridge holds a B.S. in nursing from Wayne State University and an M.S. in nursing education from the University of Chicago. Her post masters study was carried on at Ohio State University. She was coauthor with Betty J. Thomas and Mary L. Gieger of "Nurse Action in Darvon Administration," a paper presented at the clinical session of the American Nurses' Association Biennial Convention (1963).

M. Jean Daubenmire is associate professor of nursing at Ohio State University. She earned both her B.S. and M.S. in nursing and has done post masters work in education and business administration at Ohio State. Professor Daubenmire is currently completing her fourth year as principal investigator of a federally funded research grant studying "the nurse-patient-physician communicative interaction process." She has written numerous papers on nursing, nursing education, communication and research.

Kenyon B. De Greene is associate professor of human factors at the Institute of Safety and Systems Management, University of Southern California.

Previously, he taught at the University of Montana and the University of California, and has also held positions in systems analysis, design, and management at the RAND Corporation, Systems Development Corporation, Northrop Corporation, and Aerospace Corporation, where he had extensive experience in the development of large-scale systems. Dr. De Greene received his Ph.D. in physiological psychology from the University of California at Los Angeles. He is editor and principal contributing author of *Systems Psychology* (McGraw-Hill, 1970) and the author of *Sociotechnical Systems* (Prentice-Hall, 1973).

Gordon Hearn is dean of the School of Social Work at Portland State University. Dr. Hearn received his Ph.D. in group psychology from the Massachusetts Institute of Technology, his M.S. in group work education from George Williams College, and his A.B. from the University of Manitoba. He was formerly professor of social welfare and assistant dean of students at the University of California at Berkeley. Professor Hearn is the author of *Theory Building in Social Work,* published in 1958 by the University of Toronto Press, and editor of *The General Systems Approach: Contributions toward an Holistic Conception of Social Work* (New York: Council on Social Work Education, 1969). His professional affiliations include membership in the National Association of Social Workers, the American Sociological Association, and the American Group Psychotherapy Association. In addition, he is a Fellow of the American Psychological Association and of the National Training Laboratories, Institute of Applied Behavioral Science.

Joseph W. Hess is professor and chairman of the department of family medicine and professor of medicine at Wayne State University School of Medicine. He received his M.D. from the University of Utah. Dr. Hess has authored and coauthored numerous medical articles. His professional affiliations include the American College of Physicians, the American Rheumatism Association, the American Educational Research Association, and the Association of American Medical Colleges. In addition, he is a consultant on medical education for the World Health Organization and is a former member of the national review committee of the Regional Medical Programs.

Daniel Howland is professor of management in the College of Administrative Science at Ohio State University, where he also received his Ph.D. in psychology and industrial engineering. His major research interest is health care delivery systems, and he has authored and coauthored numerous articles dealing with cybernetic modeling of adaptive systems. Dr. Howland's professional affiliations include the Academy of Management, the American

Society for Cybernetics, the American Psychological Association, the American Public Health Association, the Human Factors Society of America, the Institute of Management Science, the New York Academy of Science, the Operations Research Society of America, and the Society for General Systems Research.

Richard C. Jelinek is president of Medicus Systems Corporation. Dr. Jelinek earned his Ph.D. at the University of Michigan, where he previously served as director of Systems Engineering Group, Bureau of Hospital Administration, and associate professor of industrial engineering and hospital administration. He is a member of a number of scientific and professional societies, has published many articles in professional journals, and has recently coauthored a book on systems analysis in the management of patient care (Jelinek, R. C., Munson, F., and Smith, R. L. *Service Unit Management: An Organizational Approach to Improved Patient Care.* Battle Creek, Michigan: W. K. Kellogg Foundation, 1971).

Imogene M. King is professor, graduate education, medical and surgical nursing at Loyola University of Chicago. Dr. King received her Ed.D. from Teachers College, Columbia University, and both her B.S. and M.S. degrees in nursing education and administration from St. Louis University. She is a member of several honorary and professional societies. Dr. King's major research interests in nursing are theory development and testing theoretical constructs in research in clinical nursing. In addition to a book, *Toward a Theory for Nursing* (New York: Wiley, 1971), she has also written a number of articles dealing with nursing theory, education, and research. The proceedings of the 1972 semiannual meeting of the Operations Research Society of America reports her keynote speech titled *Health Operations Research—Patient Aspects.*

George J. Klir is a professor at the School of Advanced Technology of the State University of New York at Binghamton. He received the M.S. degree in electrical engineering from the Prague Institute of Technology in 1957, and the Ph.D. degree in computer science from the Czechoslovak Academy of Sciences in 1964. He is also a graduate of the IBM Systems Research Institute in New York. Before joining the State University of New York, Dr. Klir had been with the Computer Research Institute and Charles University in Prague, the University of Baghdad, the University of California at Los Angeles, and Fairleigh Dickinson University in New Jersey; he has also worked part time for IBM and Bell Laboratories, and taught summer courses at the University of Colorado, Rutgers University, and Portland State University in Oregon. Dr. Klir's main research activities have been in general systems methodology, logic design and computer architecture, switching and automata theory, discrete mathematics and the philosophy of science. Dr. Klir is the author or coauthor of eight books and many articles, and holds a number of patents. He is a senior member of IEEE, and a member of ACM, ASC, and SGSR.

Maura Kolins is currently employed as a social science research analyst in the division of health insurance studies, Office of Research and Statistics of the Social Security Administration. Ms. Kolins has worked primarily in the areas of utilization review, organization of primary care services, and community health, with particular emphases on program evaluation and health services research.

Robert D. LaRue is professor of engineering graphics and computer and information science at Ohio State University. He holds a B.S. and an M.S. in mechanical engineering from the University of Idaho. Professor LaRue's activities and interests include computer applications, his special interests being computer graphics and numerical methods.

Mihajlo D. Mesarovic is professor of engineering and director of the Systems Research Center at Case Western Reserve University. Dr. Mesarovic holds a Ph.D. from the Serbian Academy of Science. His professional interest is in the area of systems theory and its applications to large-scale problems in industry and the biological and organizational sciences. Dr. Mesarovic has published many papers, edited and written books in the area of systems, and served as consultant for a number of well-known American companies. He holds patents in magnetic amplifiers and large-scale power systems control area in Germany, Japan, Switzerland, and Sweden. Dr. Mesarovic is a member of a number of professional societies.

Ruben Meyer is professor and program director of maternal and child health at the School of Public Health, University of Michigan. He is consultant in pediatrics for the child development project of the University of Michigan, consultant for Children's Hospital of Michigan, Sinai Hospital, and Mt. Carmel-Mercy Medical Center in Detroit. Dr. Meyer received his M.D. from Wayne State University, and has coauthored numerous medical articles. He is a member of the Governor of Michigan's State Health Planning Advisory Council. Some of his professional affiliations include Alpha Omega Honorary Society, the American Academy of Pediatrics, and the Midwest Society for Pediatric Research.

Joyce Y. Passos, now on leave from the College of Nursing at Wayne State University where for many years she was professor and chairman of the department of medical-surgical nursing and later acting associate dean, holds a Ph.D. in higher education from Michigan State University, a B.S.

from Simmons College, and an M.S. in the area of teaching medical-surgical nursing from Boston University School of Nursing. Dr. Passos has written articles concerning clinical nursing and nursing education and has participated in numerous professional programs. She belongs to a number of professional organizations, and has been active in the Michigan Nurses' Association.

Lillian M. Pierce is professor in the School of Nursing and director of nursing research at Capital University. Dr. Pierce received her M.A. and Ph.D. in behavioral science from Ohio State University, having previously earned a B.S.Ed. in biological sciences from Ohio University. She is a member of a number of honorary and professional societies. Dr. Pierce's major research interest is the systems approach to the study of patient care and health care systems; she has presented many papers and published many articles dealing with these topics.

Nicholas D. Rizzo is director of the Court Clinic in Lawrence District Court, Lawrence, Massachusetts. In addition, he is a psychiatric consultant for Phillips Academy. Dr. Rizzo holds an Ed.D. from Harvard University and an M.D. from Boston University School of Medicine. His early research interests concerned problems in the psychology of perception and learning handicaps. His medical experience began with pediatric medicine, and eventually he became a psychiatric specialist with special interest in psychosomatic medicine, adolescent medicine, and, more recently, general systems theory. He is coeditor, with William Gray and F. J. Duhl, of *General Systems Theory and Psychiatry* (Boston: Little, Brown, 1969), and with William Gray of *Unity Through Diversity* (New York: Gordon & Breach, 1973), a Festschrift for Ludwig von Bertalanffy. He is also on the consulting staff for psychiatry at the New England Baptist Hospital in Boston.

Hjalmar Rosen is professor of psychology at Wayne State University. He holds a Ph.D. in psychology from the University of Minnesota, where he also earned B.A. and M.A. degrees in psychology. His professional affiliations include the American Psychological Association, the American Sociological Association, the Industrial Relations Research Association, and the Midwestern Psychological Association. Dr. Rosen has authored and coauthored a number of books and articles in the field of psychology.

Steven H. Schwartz, associate professor and chairman of the psychology department at the University of Massachusetts/Boston, received both his Ph.D. and his A.M. in experimental psychology from the University of Illinois; he earned his B.S. in psychology from City College of New york. Before joining the faculty at the University of Massachusetts/Boston, Dr.

Schwartz was assistant professor of psychology at Wayne State University. He has authored and coauthored a number of articles in the fields of cognitive psychology and medical decision making.

Grayce M. Sills is a professor in the school of nursing at Ohio State University. She holds the B.S.N. degree from the University of Dayton and earned both a masters and a Ph.D. in sociology at Ohio State University. Dr. Sills is a member of Alpha Kappa Delta and Sigma Theta Tau, honorary societies, as well as the American Nurses' Association, the Ohio Nurses' Association, the American Sociological Association, and the American Academy of Political and Social Science. Her major research interests include methodology, role theory, and social systems. Dr. Sills has written and presented a number of papers on roles and social systems.

Roger I. Simon is associate professor of educational theory at The Ontario Institute for Studies in Education in Toronto, Canada. After receiving his Ph.D. in administrative science from Yale University, Dr. Simon taught in the department of psychology at Wayne State University. He has published articles in the areas of educational program development, complex gaming, problem solving and medical decision making, and forensic psychology.

Shirley A. Smoyak is professor, chairperson of the department of psychiatric nursing, and director of the graduate program in psychiatric nursing at the College of Nursing, Rutgers—The State University of New Jersey. She holds an adjunct appointment as professor in the department of psychiatry, College of Medicine and Dentistry of New Jersey in New Brunswick, New Jersey. Dr. Smoyak earned her Ph.D. in sociology from Rutgers, where she also received her B.S. and M.S. degrees. She serves as chairperson of the research fund committee of Sigma Theta Tau, and is a member of the editorial board of *Image*. In addition, she is a member of the American Nurses' Association's Council of Nurse Researchers and Council of Advanced Practitioners in Psychiatric Nursing. She is a charter member of the American Academy of Nursing.

James B. Tenney is assistant professor of medical care and hospitals at The Johns Hopkins University School of Hygiene and Public Health. Dr. Tenney graduated from Oberlin College, received his M.D. from Hahnemann Medical College, and his Dr. P.H. from Johns Hopkins; he is certified by the American Board of Preventive Medicine. In addition to his teaching and research activities, Dr. Tenney currently serves as a consultant on information aspects of health services systems to the National Center for Health Statistics and to the National Center for Health Services Research and

Development. He is a member of the American Association of University Professors, the American Public Health Association, and the continuing education committee of the Maryland Regional Medical Plan.

Thomas Weisman is a research associate in the Systems Division of the Graduate School of Engineering at Case Western Reserve University. Dr. Weisman received his B.A. degree from Yale University in 1968, his M.S. degree in health sciences education at Case Western in 1973, and his M.D. at Case Western in 1974. He is currently enrolled as a Ph.D. candidate in the Systems Division of the Graduate School of Engineering at Case Western Reserve, and is taking postgraduate training in internal medicine at Mount Sinai Hospital, in Cleveland, Ohio. He is the author of *Drug Abuse and Drug Counseling: A Case Approach* (The Press of Case Western Reserve University, 1972).

A. Dawn Zagornik was director of the health clinician project at Wayne State University from 1969 to 1974, and associate professor in the College of Nursing at the University. She received her M.S.N. from Wayne State University. Ms. Zagornik belongs to Sigma Theta Tau, the National League for Nursing, and the American Nurses' Association.

Health Research:
The Systems Approach

PART ONE

General Systems Theory

The initial group of papers in this volume is intended to provide an introduction to general systems theory. In addition, these papers serve as an integrating framework around which to organize subsequent papers that attempt to apply systems theory to health research. Von Bertalanffy's introduction represents a classic statement of the rationale and concepts of general systems theory. The need for a theory of "universal principles applying to systems in general" is discussed in this paper, as well as the basic principles of general systems theory and its meaning and aims. One of the major functions of the theory is the unification of science, and the consequent integration in scientific education. A number of quotations from authors not themselves engaged in the development of the theory are used to support the author's belief that men trained to practice science—"scientific generalists"—are essential to a better understanding of reality. It is important to note that underlying von Bertalanffy's well-developed theory there is a basic belief in man as an individual from whose mind stems the basic values of humanity, and he believes that continued development of the theory will conclusively demonstrate the significance of the individual mind.

The paper by Rizzo identifies many disciplines for which

1

general systems theory is relevant. He presents the theory as being very comprehensive in scope, and then identifies some specific areas of application in the health services. The many derivations from general systems theory now widely used in health fields have made a significant impact. However, the nature of their use and the people who use them are crucial factors to be considered. To say that the "focal care subsystem must be the patient" is not to be questioned, but it is vital to determine how systems theory can be utilized to guarantee this desired focus. In his discussion of the application of the theory, Rizzo gives particular consideration to psychology and psychiatry. He proposes that conceptual models based on general systems theory be used to explain the human mind because these models correspond more closely to reality than previously used reductionistic frames of reference. His plea for the use of theory by physicians might well be applied to all other health professionals. He suggests that the enticing pragmatic approach, which produces immediate results, can result in a technical discipline masquerading as a profession.

In the final paper of this introductory section, Hearn expresses some dismay with the fractionalization and specialization within the social work discipline, on the premise that they do not permit the development of a theoretical model of the client system which can contribute to an understanding of the reality of the client's life. Hearn expresses a need for a basic conceptualization of the social work process, a conceptualization that is applicable whether the client is an individual, dyad, small group, or larger social system. He goes on to describe a model that includes the concepts of system universe, client subsystem, change agent, and the means by which systems maintain their viability. The human system model is described in terms of subsystems, e.g., the energy subsystem; the value subsystem; and positional, role, and relational subsystems. Clearly, both models view the client as the focal subsystem. Such models could be the basis for intervention within any discipline. Their potential significance is now limited primarily by their stage of development, predictiveness, and comprehensiveness.

An understanding of the content of the papers in this section will permit the reader to move from a comprehension of the abstract concepts of general systems theory to the analysis of applications of those concepts in various intervention sub-

systems. The remainder of this volume is devoted to the presentation of the applicability of systems concepts to health research.

CHAPTER 1

Introduction

Ludwig von Bertalanffy

Dr. Harriet Werley was kind enough to ask me to introduce the present symposium, which, because of illness, I was unable to attend. According to Dr. Werley, "What we have in mind is a general statement of systems theory and concepts specifically geared to the health specialist."

A considerable number of publications attest to the vivid interest taken by the health services and their professionals in general systems theory (Crosby, 1971; Hazzard, 1971; Hodges, 1971; Kinney, 1971; Miller, 1971; Moughton, 1971; Reres, 1971; Sheldon, 1970). Perhaps we can best introduce the present book if we try to answer the three-part question — why has this interest developed; what may the "systems approach" offer to the field, and how may the suggestions so obtained be implemented?

If we compare the general practitioner or family doctor of old with the workings of a large modern hospital and the complex apparatus of health care, the differences are in many ways similar to those between an old-fashioned craftsman and a large industrial firm. What works well for the individual artisan and his customers is not the same as what applies to a large industry and an anonymous mass of prospective buyers.

It is for these reasons that industrial, commercial, military, and similar enterprises have found it necessary to adopt new methods and techniques, such as the wide use of computers, organizations research, systems analysis, linear programming, and operations research. The advantages and disadvantages of these new methods and techniques are apparent. Production on the assembly line caters to a mass of people that previously was excluded from the luxuries and even the necessities of life; it is apt, on the other hand, to reduce the quality of the product and to menace the satisfactory life both of worker and consumer.

This, in part, is a direct consequence of growth, size, and the laws governing human enterprises. An airplane model enlarged to the size of a jet plane could, for well-defined technical reasons, never fly, nor could an elephant-sized mouse survive. The same is true of human organizations. The dialectical principle of the transformation of quantity into quality applies; that is, quantitative changes, an increase in size, transmute into qualitative changes — particularly in the case of the transition to higher levels of organization. Thus the structure of the university, for example, changes in various — not altogether favorable — ways by becoming the modern, giant "multiversity" (Gallant and Prothero, 1972). Something similar is true of social change in general.

Expressed in a somewhat different way, things have become so complex in modern society — and, in other respects, in modern science, too — that the conventional procedures of old are no longer sufficient, and new ones are required. Speaking in somewhat more technical language, it is essentially with problems of "many variables" or "system problems" that we have to deal.

In consequence, a new discipline, or rather a bundle of disciplines, has developed, which can be summarized under the title of "general systems theory": developments that emerged both in science and in its technological, industrial, and commercial applications. I shall not, in these introductory remarks, enter into a summary of these developments but may refer the reader in the health sciences to a review that Dr. Frank Baker of the Massachusetts Institute of Technology gave in the book, *Systems and Medical Care* (1970); those desiring more detailed information will find this in recent works of which perhaps the present author's *General System Theory* (von Bertalanffy, 1968) and the books edited by G. Klir (1972); by W. Gray, F. D. Duhl, and N. D. Rizzo (1969); and by W. Buckley (1967) may be found most useful.

Now the principles obtaining for large organizations obviously hold true also for a large hospital or for health services in general — they, too, are very complex "systems" and therefore, in some way or other, have to use the systems approach.

This roughly delineates the justification of the present book and the symposium upon which it was based. The argument appears to be rather straightforward and evident. Immediately, however, we see a host of problems that stem from the special case with which we are concerned.

It would be hard to deny that present health service in the United States, comprising the doctor's practice, hospital service, and health care in general, is in a sorry state at any of these levels, notwithstanding the progress made in medicine during the last few decades. Many books, or rather libraries, of criticism have been written on this topic, and detailed discussion would far exceed the present introduction. Instead, it may perhaps suffice to quote a few newspaper headlines and the like that can serve as an arbitrary but characteristic "sampling" of present complaints.

The assumption that effective treatment of illness vastly improves community health is incorrect . . . . Most of the decline in disease incidence and mortality, and therefore most of the increase in average life expectancy, has resulted from influences other than efforts aimed at controlling specific diseases . . . our obsession with reductionism has led us to ignore the very real values of a synthetic systems-oriented approach (Stallones, 1972).

During the critical period when most men are at the peak of their working and earning capacity (i.e., between 45 and 55) twice as many die in the United States as in Sweden — a mortality rate that is also higher than that of almost every Western nation (Field, 1970, p.146).

We die sooner, pay more, and East Germany does better by the babies (Headline in *New York Times Book Review*, March 7, 1971, p.3).

Uneeded Surgery Kills 10,000 a Year, Aide to Nader Asserts (Headline in *Buffalo Evening News*, December 18, 1971, p.7).

Pharmaceutical firms were spending about $5,500 for every doctor in the U.S. to promote the sale of their products, employing 27,000 detail men to persuade doctors and drugstores to prescribe and sell their particular brands. In other words, while the medical profession is protesting its integrity, the doctor-patient relation, the necessity of keeping a high standard (and income deriving therefrom which, even for the most mediocre practitioner, is a multiple of that of an academic authority), the stupendous research activity of drug companies, etc., actually it is not the *doctor* who prescribes a drug according to the patient's condition and his medical knowledge, but the *salesman* (advertising artist, etc.) who has the loudest voice in the market (possibly excepting aspirin which, in American practice, appears to function as a panacea where no advertised product comes to the doctor's mind). In still other terms, the practices of Wild West quacks, bogus patent medicines, etc., have not much changed in modern practice (von Bertalanffy, 1967, p.121).

The American hospital up from the English pest hole, was and is a hotel for the doctor's patients who must suffer whatever indignity or inconvenience will benefit the doctor . . . . The huge health-insurance industry, some $20-billion per year in buying power, has looked upon itself as a service for the medical profession. Its interest in the consumer has been limited in the past to estimating what the public could and would buy . . . much of medicine is guess-and-by-gosh faith healing; lab tests and x-rays mean different things to different doctors; annual examinations are of dubious value; periodic examinations for rectal cancer are equally dubious; low-cholesterol diets and jogging for the weighty are not all good; and half a dozen surgical operations, including tonsillectomies, are questionable as therapy (*New York Times Book Review*, January 23, 1972, p.5).

Speaking in more general terms: the United States boasts of being the most affluent nation in the world, having the highest Gross National Product; the country provides all sorts of luxuries by creating artificial needs that, at the same time, lead to waste of resources and pollution of the environment — but it lags far behind other countries in the quality of medical care generally available, in the doctor-patient ratio, the increase in life span (the U.S. holds 19th place in this scale), health care for the indigent, and in many other respects. Medical spectaculars and stories of "breakthroughs" in modern medicine appearing in *Time* magazine and in the other media—such as that of a host of doctors, at vast expense, saving the life of one little girl (for how long?) with the use of some sensational technique or machine—cannot obscure the fact that treatment of the common and medically banal diseases, such as heart disease or appendicitis, often falls short of elementary standards. If a doctor when faced with a case of appendicitis prescribes castor oil or operates for an ovarian cyst (cases of which I have personal knowledge), and if malpractice suits have reached an all-time high, the inference cannot be avoided that the health service (as many others, from the mails to public transportation and the university) is deeply sick. The advertising techniques of pharmaceutical companies not only in television commercials but in respectable medical journals are no different from those for patent medicines by nineteenth-century quacks and, for that matter, of Dr. Dulcamara in Donizetti's *L'Elisir d'Amore.*

These facts are generally known but unwillingly heard. Much more could be said, but it is perhaps well to remember that health insurance was introduced in Germany by the archreactionary Bismarck, while in the "democratic" United States it took some 70 years more to develop because of resistance against "socialized medicine"; and it eventually arose as a chaos of agencies (Blue Cross, Blue Shield, Medicare, Medicaid, etc.) whose appropriate reorganization and coordination most definitely needs a "systems approach"—the more so as health care has become a political football.

But there is more involved that makes our initial comparison between industry and health care, though valid to a point, inadequate in others. According to this scheme—which is to a large extent realized in practice—the hospital becomes a sort of health factory, with sick people as input and supposedly cured ones (plus cadavers) as output, just as, in an automobile plant, iron and other raw materials go in at one end and finished cars come out at the other. But such a "model" becomes most disquieting when our concern is not with industrial products but with living and suffering human beings. The consequence is the general complaint that medical service and health care become progressively depersonalized and dehumanized, (with the exception of that given by some nice nurses occasionally encountered), and the hospital grows into a sort of prison; with the difference, characteristically, that more thought and publicity seems to be given to the complaints of prison inmates than to the feelings of hospital patients.

This, in part, is a consequence of increasing numbers and of the "bigness" to which we have already alluded. The doctor can "relate" to a limited number of patients; beyond this, they become an anonymous and

faceless mass. Diagnostically, the patient becomes a bundle of tests; thera-
peutically, a file card with orders to the nurse. What is lost in the process is
what in former times was called the "clinical vision" of the practitioner, that is
the experienced doctor's ability to *see* what is wrong instead of relying on a
battery of tests and a host of specialists, and that "doctor-patient relation,"
the immeasurable psychological transaction that is part of any, even the most
"somatic" or surgical, treatment. The loss of this Hippocratic spirit is possibly
the most important consequence of the "hospital-as-factory," and of the
conversion of the practice of medicine from a "calling" into a lucrative
business. It is probably not, or only partly, compensated by the well-
advertised "breakthroughs" and "wonder drugs" of modern medicine. No one
in his senses will deny the benefits derived from modern surgery, antibiotics,
diagnostics, and so forth. But, as René Dubos has so often emphasized, one
may not forget that the increase in life span and the disappearance of the
great epidemics, are due rather to improved living conditions and hygiene
than to specific cures.

What has this to do with the "systems theory" with which we are
concerned? I think it is rather important for its evaluation.

Trivially, and according to considerations such as those just mentioned,
medical practice and health care should apply the most modern systems
methods, both in their "hardware" (computers and allied machinery) and
"software" (organizational techniques and the like). For only in this way, can
the problems resulting from "bigness" in a mass society, and at the levels of
the individual doctor, of the hospital, and of health insurance, be handled
more satisfactorily than if they are left to the chance availability of doctors,
hospitals, and health plans (or conversely, to the competition of patients for
acceptance by doctor and hospital). "Organization" or "planning" in this
sense is unavoidable. "Free competition" in its nineteenth century sense has
long been abandoned in industry and commerce and only subsists, to any
great extent, in the A.M.A. and in the field of health care. The ways to deal
with such problems are highly developed and must be elaborated under
proper directives for special cases.

Here it should be remembered that there are two main trends in the
systems approach, which, for short, we may call the "mechanistic" and the
"organismic-humanistic."

The mechanistic approach emphasizes, theoretically, cybernetics,
regulation by feedback, computerization, and the methods of systems engineering as used in business and industry. While acknowledging the usefulness of flow charts, benefit-cost analyses, and the like for the improvement of
health services, the present writer does not share the enthusiasm of those who
believe that the computer will automatically lead to a solution of the many
problems of the present — those of diagnosis, medical care, and administration in particular.

Speaking as a private individual, I have not felt the benefits of computerization. What one does experience are the many inconveniences arising from all-too-frequent computer errors, a chaos of innumerable "forms," long delay in payment of claims supposedly covered by health insurance agencies, aggravation of one's tax problems, a flood of unwanted junk mail, and the like. One is inclined to ask to what extent this is to be ascribed to the volume of business arising in a mass society, or to computerization running wild, as it were, when the old-fashioned secretary would do a job that is more limited to the necessary activities, but more efficient. While computers have made interplanetary flights possible, paychecks sent to the wrong address, health insurance payments indefinitely delayed, erroneous charges by American Express, muddled up students' registrations, and a thousand similar trivia do not seem to show that the computer has made life easier. On the other hand, fears that a "second industrial revolution," brought about by the replacement of human labor with automation, would lead to previously unheard-of unemployment and problems having to do with the utilization of leisure time seem to have been exaggerated. Our present worries — pollution, drug abuse, busing, crime, the population explosion, the dollar crisis, war, and politics — have much to do with industrialized mass society but little with the "computer revolution" in the sense in which the founders of cybernetics had predicted.

Still speaking from personal academic experience, we have heard complaints about the "depersonalization" of the university, the alienation between faculty and students, the students' riots of past years, all of which have profoundly endangered the status of the university and are a consequence of its "bigness." But, in my own teaching, I had classes of eight hundred or a thousand students some thirty years ago, with no one feeling "alienated" or confronted with the now-current psychological problems — an experience which, I believe, forcefully applies to medical education and the training of paramedical personnel.

As to the medical and clinical application of the computer, I may quote a respectable authority in the field. Dr. Irwin D. Bross (1971), Director of Biostatistics at the well-known Roswell Park Memorial Institute in Buffalo, writes:

For the past 20 years the mass media and the scientific and technical journals have been filled with glowing accounts of all the marvelous things that electronic computers were going to do for science. But when we take a close look at what these so-called "giant brains" have actually accomplished in the sciences — or in biomedical research in particular — it is hard to name a single major discovery or advance in which computers played any essential part.

All sorts of glamorous uses of computers have been highly publicized — computerized diagnosis and computerized hospital operation, for instance. But if one takes the trouble to read the whole story, it usually has an unhappy ending. About five years ago, a whole series of computerized hospitals were announced with great fanfare. But despite a flurry of enthusiastic preliminary reports in *Science* and other supposedly reputable journals, every single one of these systems has been junked. There are a few bright exceptions — notably in very prosaic and unglamorous jobs. But I believe that it is fair to say: *So far computers have been a big flop* (p.4).

I am not competent to discuss the merit of Dr. Bross' argument. Rather I would like to call attention to what may be denominated the "humanistic" trend in systems theory. It duly acknowledges and makes use of all theoretical and practical approaches that the "mechanistic" trend legitimately puts forward but transcends it by what W. Gray (1972) honorifically called the "five Bertalanffian principles." These are: (1) the organismic systems or non-reductionist approach, emphasizing the wholeness of the organism and its accessibility to scientific research, contrasted with the elementaristic and summative approach of conventional science; (2) the principle of the active organism in contradistinction to the reactive organism, the robot or S-R scheme; (3) the emphasis on the specificities of human compared with animal psychology and behavior, subsumed under the notion of symbolic activities; (4) the principle of anamorphosis, that is, the trend toward higher order or organization — in contrast to the entropic trend in ordinary physical processes — which is made possible by the "open-system" nature of the living organism and is expressed in "creativity" and its manifold manifestations ranging from evolution in its nonutilitarian aspects to behavior in play and exploratory activities to the highest human creativity and culture; and (5) as consequence of the latter, the introduction of specifically human and supra-biological values into the scientific world view.

The impact of this orientation on medical theory and practice has been discussed in various places. In the present context I would like to mention only one of its consequences. The contrast to the conception of the doctor as a glorified auto mechanic — replacing damaged parts by artificial organs or, in the futuristic expectations of "genetic engineering," replacing damaged genes and manufacturing clones of identical individuals — and to the hospital as a health factory, is obvious.

This is particularly confirmed by the strong influence general system theory has exerted in modern psychiatry (W. Gray, F. J. Duhl, and N. D. Rizzo, 1969; K. Menninger, M. Mayman, and P. Pruyser, 1963; R. Grinker, 1967; S. Arieti, 1967; L. von Bertalanffy, 1966).

With respect to health care, it is not the intention of this introduction to propose any specific measures (as could well be done) because this is the main topic for the papers to follow. But we are entitled to say that the systems approach in its various aspects will be indispensable if better solutions of pressing problems are to be arrived at. And, in its "humanistic" trend, it may contribute toward a reorientation of medical thought and attitude. Let us not forget what Hippocrates already knew some twenty-five centuries ago: that against the trend toward depersonalization, human values such as kindness, loving care, and personal relationships form an essential part of health care.

The manifold contributions contained in the present volume and united by the thread of system thinking will present many different aspects of a pressing problem and, hopefully, contribute toward an alleviation of the crisis in modern medicine to which we alluded in the beginning.

REFERENCES

Arieti, S. *The Intrapsychic Self*. New York: Basic Books, 1967.

Baker, F. General systems theory, research, and medical care. In A. Sheldon, F. Baker, and C. McLaughlin (eds.), *Systems and Medical Care*. Cambridge, Mass.: M.I.T. Press, 1970, 1-26.

Bertalanffy, L. von. General system theory and psychiatry. In S. Arieti (ed.), *American Handbook of Psychiatry*, Vol. 3. New York: Basic Books, 1966. pp. 705-721.

——. *Robots, Men and Minds*. New York: Braziller, 1967.

——. *General System Theory: Foundations, Development, Applications*. New York: Braziller, 1968.

Bross, D. J. Professor says computers have been flops in science. *Reporter* (State University of New York at Buffalo), February 18, 1971.

Buckley, W. (ed.). *Sociology and Modern Systems Theory*. Englewood Cliffs, N. J.: Prentice-Hall, 1967.

Crosby, M. H. Control systems and children with lymphoblastic leukemia. *Nursing Clinics of North America*, 1971, *6*(3), 407-313.

Field, M. G. The medical system and industrial society. In A. Sheldon, F. Baker, and C. McLaughlin (ed.), *Systems and Medical Care*. Cambridge, Mass.: M.I.T. Press, 1970, 143-181.

Gallant, J. A., and Prothero, J. W. Weight-watching at the university: The consequences of growth. *Science*, 1972, *175*, 381-388.

Gray, W. Bertalanffian principles as a basis for humanistic psychiatry. In E. Laszlo (ed.), *The Relevance of General Systems Theory*. New York: Braziller, 1972, 123-133.

Gray, W., Duhl, F. J., and Rizzo, N. D. (eds.). *General Systems Theory and Psychiatry*. Boston: Little Brown, 1969.

Grinker, R. (ed.) *Toward a Unified Theory of Human Behavior*. Second edition. New York: Basic Books, 1967.

Hazzard, M. E. An overview of systems theory. *Nursing Clinics of North America*, 1971, *6*(3), 385-393.

Hodges, L. C. Systems and nursing care of the cardiac surgical patient. *Nursing Clinics of North America,* 1971, *6*(3), 415–424.

Kinney, A. B., and Blount, M. Systems approach to myasthenia gravis. *Nursing Clinics of North America,* 1971, *6*(3), 435–453.

Klir, G. (ed.). *Trends in General Systems Theory.* New York: Wiley, 1972.

Menninger, K., Mayman, M., and Pruyser, P. *The Vital Balance.* New York: Viking Press, 1963.

Miller, J. C. Systems theory and family psychotherapy. *Nursing Clinics of North America,* 1971, *6*(3), 395–406.

Moughton, M. L. Systems and childhood psychosis. *Nursing Clinics of North America,* 1971, *6*(3), 424–434.

Reres, M. E. Systems analysis: An approach to working with personality disorders. *Nursing Clinics of North America,* 1971, *6*(3), 455–462.

Sheldon, A., Baker, F., and McLaughlin, C. P. (eds.). *Systems and Medical Care.* Cambridge, Mass.: M.I.T.Press, 1970.

Stallones, R. A. Community health. *Science,* 1972, *175,* 839.

CHAPTER 2

General Systems Theory: Its Impact in the Health Fields

Nicholas D. Rizzo

Why was general systems theory given a prominent place in this symposium? What does general systems theory have to do with medical care and the practical implementation of health services?

The field covered by general systems theory is a vast intellectual, conceptual, technical area that is as yet loosely circumscribed. The applications of general systems theory range from abstract mathematical equations that attempt to define a particular system to awesome, complex, self-regulating machines. In some of the disciplines for which systems theory is relevant, the immediate problems include questions and issues on the nature of man and society, pollution, overpopulation, and criminality. General systems theory offers a world view that is, in many respects, highly different from what some may expect. Perhaps most important of all, general systems theory is not an apology for the "establishment" or the "system" under which we live. Far from it! In ever-increasing numbers, societies and institutions are treating man as a machine, as a robot, as a replaceable part, as a numbered file in a medical clinic, or as a changeable working unit on an assembly line. We live in a time of rapid change in technology, communications, and travel; efficiency, speed, and materialistic gain are overriding themes. It is no idle fear that we may soon be buried in our own junk. Certain laws of national revenue and fiscal policy contribute to this sad state. Item: during the past decade it has become possible to depreciate, or write off, buildings for federal income tax purposes in ten years, in order to feed an insatiable and growing economy

15

which seems to lack a head but feeds more and more on a diet of inflation, expansion, and planned obsolescence.

General systems theory as a concept was formulated by von Bertalanffy nearly 40 years ago (Gray and Rizzo, 1969). It must be kept in mind that von Bertalanffy, first a biologist, developed his theories partly because of his dissatisfaction with the atomistic, mechanistic views of biology prevalent a half-century ago. Similar dissatisfactions were occurring among psychologists, physicists, and more recently sociologists and economists. The major contribution of von Bertalanffy was to restore to biology the concepts of unity, wholeness, and organization. If one believes in a universe of randomness, or chance, or happenstance (Blandino, 1969, 1973) he will very likely assign values of importance to human beings far different from the values that are apt to result from a holistic approach to the major problems of biology. The search for unity among thinkers, philosophers, and teachers can be traced in the writings and thoughts attributed to Empedocles, Heraclitus, St. Augustine, St. Thomas Aquinas, Nicholas of Cusa, and Giambatista Vico (Durant, 1939, 1944, 1950; von Bertalanffy, 1971). In the field of psychology, Grinker has written of a third revolutionary change brought about by general systems theory (von Bertalanffy, 1967), the earlier two being the advent of psychoanalytic theory and behavioristic psychology. The psychology which is referred to as holistic, organismic, gestalt, or configurational did not begin with general systems theory; it arrived on the scene earlier, as is evident in the writings of four German psychologists, Koehler, Koffka, Metzger, and Wertheimer (Rizzo, 1972a). They were later joined by an American, Raymond H. Wheeler (1932), who held views similar to those of von Bertalanffy, particularly as he developed them in *Robots, Men and Minds* (1967). The hierarchical arrangement of mental functions in man and the human capacity for spontaneous activity, rather than the ability to react when stimulated, are two of the chief features of the psychology and physiology derived from organismic humanistic biology (Francoeur, 1970). Differentiation of some functions into more refined specific skills, from a more generalized ability, is another feature. Man grows and develops by a process of differentiation, refinement, and individuation; a machine or robot grows by having parts and therefore new functions added. The psychological principles referred to, those of wholeness and unity, actually the cornerstones of organismic psychology, illustrate a striking parallelism to the organismic biology of von Bertalanffy.

The basic principles of general systems theory have been derived from two historical roots, the ancient metaphysical, psychological, and philosophical source and the theoretical biological approach which was given impetus by von Bertalanffy during the past half-century. The contributions of von Bertalanffy were, however, more than armchair theorizing; his contributions and the range of disciplines involved encompass aspects of all the so-called basic sciences of medicine: physiology, morphology, anatomy, cellular physiology, metabolism, and excitation. Even the classical sciences of mathematics and physics are involved in systems thinking. A clear distinction must be made, however, between the models derived from systems engineering, or cybernetics, and those derived from systems which are classified as open, living, and general. The former stress efficiency in process while the latter strive to maintain a humanistic bias.

SYSTEMS AND HUMAN VALUES IN MEDICINE

Many derivatives of general systems theory are represented by the topics in this symposium — organization systems, computer systems, cybernetics, health systems models, operations research, and information theory, to mention a few. All may play an important role in many health fields, or they may not, depending on how they are used, who uses them, and also why. The organization of medical services for large populations has encountered many practical obstacles (Sheldon, Baker, and McLaughlin, 1970). Many sectors of organized medicine have only recently acknowledged the fact that there has been a shortage of physicians in this country for several years. In those instances where hardware, no matter how sophisticated, is used to replace a doctor in the diagnosis and management of sick people, the practice of medicine becomes dehumanized and not worthy of its lofty heritage. Can such really happen? Unfortunately, it can and does happen. We have available today an automatic scanning device for the early recognition of cancer cells, devices for the mechanized monitoring of patients with some forms of cardiac disease, and mechanical methods for the evaluation of electrolyte imbalance.

It cannot be emphasized too strongly nor too often that the focal care subsystem must be the patient, the human being, as Dr. Hearn puts it in the title of his presentation. The patient is not often enough the center of concern.

The public image of the physician gained from mass media does not correspond with realities encountered by many citizens. In summarizing the above remarks, it is necessary to recall that the traditional model of the medical profession, for physicians as well as nurses, was one person ministering to a patient in need.

The reader should acquaint himself, especially from the patient's point of view, with the problems in obtaining good medical care. It is a rare community that has in operation a comprehensive medical plan that covers all its citizens. Until a few years ago organized medicine did not accept the premise that competent medical care was every person's birthright. Health facilities, covering all the traditional specialities, and the social agencies necessary to meet today's needs in most communities have not usually been carefully planned out in advance, but rather added on one to another. It behooves all health professionals, whether they be in positions of planning jobs or of implementing health goals, to develop a lively awareness of the many parallel systems at work.

The Commonwealth of Massachusetts, through its Division of Legal Medicine of the Department of Mental Health, has established a series of psychiatric clinics in courts. The principal activity of these outpatient facilities is the diagnosis and treatment of emotionally disturbed offender-patients. During the preparation of a recent report on a court clinic's work (Rizzo, 1972b, 1973) it became evident that many social agencies, institutions, and organizations, whose general aims were to provide cultural, educational, and health services for its younger citizens, were primarily isolated power bases interested chiefly in self-perpetuation and aggrandizement which they would not jeopardise by stretching a point here and there to accommodate a youngster in need. One example which is unfortunate and only slightly amusing: In the jurisdictional area of a district court there was built several years ago a modern, fully equipped, vocational and technical high school. Originally the vocational high school was intended to meet the educational needs of boys and girls who had low interest, aptitude, and success in the conventional high school with its usual emphasis on college preparatory subjects. At the same time that each pupil studied metal and woodworking, culinary arts, and automotive repairs in order to become equipped for entering one of a large number of trades and occupations, the more traditional subjects, specifically English, mathematics and social studies, were deemphasized but not eliminated. The point of such an admission policy was to make the education and training a continuing process for that group of adolescents who were not able to succeed in each of the conventional high schools. Now there is a huge waiting list, perhaps equal in number to the active enrollment. Preference is currently given to applicants with average or above average IQ scores and with passing grades or better in English, mathematics, and social studies. The school denies admission to the very pupils it was set up to serve!

GENERAL SYSTEMS THEORY IN
BIOLOGY, MEDICINE, AND PSYCHOLOGY

The publication of von Bertalanffy's *Modern Theories of Development: An Introduction to Theoretical Biology* (1933) marked the beginning of a new era in dynamic science. Von Bertalanffy wrote of "wholeness," "organization", and "teleology" at a time when physics reigned as the only indisputable science, with mathematics as its chief support. A decade previously, the four prominent German psychologists mentioned earlier, each of whom had a solid grasp of human physiology, wrote of comparable principles in psychological terms. The development of general systems theory in von Bertalanffy's publications is dramatic for several specific reasons. It is evident that he was searching for a global view from the outset, in the tradition of great scholars of the past. Biology, the science of living organisms, was his starting focus, the general field that occupied his interest. From general biology he branched out into physiology, growth and metabolism, tissue respiration, and cytology, especially exfoliative cytology. The biological basis of cytological cancer studies is being continued both here and abroad with more than 300 published scientific works of special interest to surgeons, as stated by Arminski (1973). In physiology the traditional concept of homeostasis has been replaced by the term "steady state." General systems theory has been the unifying schema in numerous symposia devoted to health care.

In the 27-year interval between the publication date of von Bertalanffy's epoch-making *Modern Theories of Development* and the founding of the Society for General Systems Research in 1955 (by von Bertalanffy, theoretical biologist; Anatol Rapoport, psychologist and mathematician; Kenneth Boulding, economist; and Ralph Gerard, physician, physiologist, and neurologist) significant articles were published by von Bertalanffy and his students on the implications and applications of general systems theory in many disciplines now considered basic in the health fields. Nearly 25 years ago the basic research in cytology led to the well-known acridine-orange screening test for detection of early cancer. The proponents of general systems theory, nearly all influenced by von Bertalanffy directly, not surprisingly turned their attention to some of the psychological problems facing the human species. In academic psychology and its derived clinical disciplines, one saw the same reductionistic frame of reference. Bits and pieces of mental life were schematically arranged in order to account for the marvelous phenomenon known as mind; this, in outline form, is the additive, hierarchical approach which is basic to behavioristic psychology.

Organismic psychology is a natural derivative of organismic biology and is not the offshoot of the static life sciences of an earlier era (von Bertalanffy, 1968). The first American psychologist to build a university department and train students in organismic psychology was Raymond H. Wheeler of Kansas, an inspiring teacher who brought Koehler, Koffka, and Wertheimer to his departmental meetings. Wheeler had studied in Germany and understood the rudiments of organismic biology. *The Laws of Human Nature* (1932), published by Wheeler four or five years after von Bertalanffy's *Modern Theories of Development* (translated 1933), mentions specifically those principles of general system theory which all students of the subject have come to regard as axiomatic and basic: functional analysis, the primacy of wholes, the principles of closure, differentiation, and individuation. A recent application of organismic psychology is Gray's (1973) subtle and potentially far-reaching concept that emotional nuances (refinements further down the road from individuation) are the organizing principles of unusual mental phenomena.

It has been a truism in this country that physicians in general and psychiatrists in particular have not been required to study psychology. Probably it is just as well, considering the nature of the material that has traditionally been taught. It must be stressed that, in contrast to behavioristic thinking, general systems theory can furnish conceptual models of mental activities as they occur in man, models that correspond more closely to the realities of the human mind than do those of mechanistic, atomistic, or behavioristic psychology. General systems theory is at the same time an excellent and rewarding research attitude and theoretical frame of reference, especially in studying problems of growth, development, acculturation, and adaptation, as well as their aberrations and vicissitudes.

GENERAL SYSTEMS THEORY IN
CLINICAL PSYCHIATRY

Within the past decade general systems theory has become the theoretical framework for an increasing number of psychiatrists, some of whom have been leaders in American psychiatry for three or four decades (von Bertalanffy, 1967, 1968, 1971). It is in the area of the interpretation of manifest symptoms that general systems theory differs from psychoanalytic theory, for example. According to general systems theory the onset of depressive, psychotic, neurotic, or psychosomatic symptoms may signal a temporary organismic breakdown rather than a simple regressive phenomenon. Published studies of court clinic patients, nearly all of whom were guilty of illegal acts (Rizzo, 1973) have yielded a number of findings. The chronic indifference of most psychiatrists toward assessing the results of treatment is an established fact. One generally recognized aim of psychoanalytic treatment, among others, is to get the patient to accept himself as he is and to develop his insights accordingly. This can lead to catastrophic results, especially when no one else can accept the patient as he is. In many instances, among court clinic patients, small behavioral changes in selected patients have been converted into major therapeutic gains (Rizzo, Gray, and Kaiser, 1969). The court clinic as a focus of community interaction has been under more or less constant study for several years and some unexpected findings have been reported. Some 310 consecutive referrals for psychiatric evaluation and treatment among boys and girls attending two private boarding schools were compared with 138 consecutive cases evaluated and treated in a court clinic. Age and sex distribution were comparable. Nearly 15 percent of those referred for psychiatric consultation and treatment from the private school population were suffering from severe psychosomatic disease or depression accompanied by suicidal preoccupation, if not gestures (Rizzo, Gray, and Kaiser, 1969). The court clinic group showed not a single case of suicidal preoccupation or gesturing and none of the children had severe psychosomatic problems.

The court clinic has offered an excellent opportunity to study an ubiquitous, necessary social system — the law — which has been characterized as an open system that frequently reaffirms its open system character by reversal, change, and modification. Explorations into the reward-punishment interfaces, criminality as an ego syntonic and tragically adapative modality, and the theoretical basis as well as practical aspects of enforced psychotherapy have yielded some important observations (Gray and Rizzo, 1973; Rizzo, 1973). Psychotherapeutic results obtained in the court clinic have been published. According to certain definite criteria, more than 65 percent of the offenders treated showed improvement.

General systems theorists have been interested for many years in the etiologic aspects of schizophrenia and have opened up a field of inquiry into the disturbed nature of the schizophrenic's use of symbols. These writings are provocative and promising. Until the advent of the phenothiazines nearly 20 years ago, etiologic considerations were discussed with as much enthusiasm and as little light as exists at present. The clinician must live by his commitment to help by all ways possible to restore at least a semblance of normal functioning to the acute schizophrenic patient. At the present writing, psychotherapy alone is not enough.

SUMMARY AND CONCLUSIONS

General systems theory is a large domain and it has assumed the dimensions of an historic movement, worldwide in scope and range. Many diverse fields have produced their own systems and theories, but not all systems are open, general, or living, though they may serve some human need. The prototype of the open, living, general system is the human organism; its inherent capacity for spontaneous activity is its unique and most valuable characteristic. Its capacity for symbolic achievement is at the core of all cultural attainments. Psychology in America has largely been a mechanistic-atomistic-behavioristic hodgepodge, and this has accounted, in part at least, for the current confusion that characterizes many disciplines derived from, or dependent upon, psychology. It is my opinion that psychology, the study of the mind, is a more important subject than most people realize. Perhaps the single most important basic element in human psychology is the principle clearly stating the primacy of "wholes." Many corollaries have been derived from this overriding rule. Growth and development are processes of refinement, differentiation, and individuation proceeding from an earlier unity. Repeated confirmations of this basic principle are found in the psychological studies of Piaget, Werner, and Bruner. But organismic or gestalt psychology has never achieved great popularity or even wide acclaim in academic circles in the United States. To believe, for example, that the pragmatic or eclectic approach is enough, without any thought being given to a theoretical or conceptual model, leads to chaos, a patchwork of meaningless examples and, sometimes, to no significance beyond the moment. Physicians, and especially psychiatrists, who are scornful of theory are, in truth, technicians pretending to be professionals. The most attractive and important aspect of the professional worker is his devotion and his sense of duty to another human being in need. No machine, in the foreseeable future, can meet that challenge.

REFERENCES

Arminski, T. C. A fluorescence microscopy method of cytologic study of rectal smears as a screening technique. Part II. In W. Gray and N. D. Rizzo (eds.), *Unity Through Diversity*. New York: Gordon & Breach, 1973, pp. 791-804.

Bertalanffy, L. von. *Modern Theories of Development: An Introduction to Theoretical Biology*. Translated and adapted by J. H. Woodger. London: Oxford University Press, 1933. Von Bertalanffy's original work, published in German, was entitled "Kritische Theorie der Formbildung" and published by Gebrüder Borntraeger in Berlin in 1928.

————. *Robots, Men, and Minds: Psychology in the Modern World*. New York: Braziller, 1967. Unfortunately, this book is too compactly written for the general reader, but it is an elegant summary of modern psychology and the dilemma facing modern man.

————. *General System Theory: Foundations, Development, Applications*. New York: Braziller, 1968.

————. System, symbol, and the image of man. In I. Galdston (ed.), *The Interface Between Psychiatry and Anthropology*. New York: Brunner/Mazel, 1971, pp. 88-119. Papers delivered at the 1969 and 1970 Academic Meetings of the American College of Psychiatrists.

Blandino, G. *Theories on the Nature of Life.* New York: Philosophical Library, 1969. Blandino is one of the leading theoretical biologists of western Europe. He is also known for his scholarly writings in Christian theology and philosophy.

Blandino, G. Chance and design. In W. Gray and N. D. Rizzo (eds.), *Unity Through Diversity.* New York: Gordon & Breach, 1973, vol. 1, pp. 377-390.

Durant, W. *The Life of Greece.* New York: Simon & Schuster, 1939.

––––––. *Caesar and Christ.* New York: Simon & Schuster, 1944.

––––––. *The Age of Faith.* New York: Simon & Schuster, 1950.

Francoeur, R. T. Man: Naked robot, or an active personality system. *Parapsychology Review*, 1970, *1*(4), 18-21. The critical review by Francoeur of two expositions of general systems theory is an eloquent statement for humanistic psychology. It should be read by all serious students of psychology, especially those who find the practical approaches of behaviorism or neobehaviorism useful.

Gray, W. Emotional cognitive structures: A general system theory of personality. General Systems, 1973, *XVIII*, 167-173.

Gray, W., and Rizzo, N. D. History and development of general system theory. In W. Gray, F. J. Duhl, and N. D. Rizzo (eds.), *General Systems Theory and Psychiatry.* Boston: Little Brown, 1969, pp. 7-31. The first textbook devoted exclusively to general systems theory and its applications to psychiatry. The first chapter gives a concise account of the history of general systems theory and parallel developments in the traditional, behavioral, and social sciences, which serve to differentiate the origins and current status of the newer systems sciences.

Gray, W., and Rizzo, N. D. (eds.). *Unity Through Diversity.* New York: Gordon & Breach, 1973, 2 vols. This work, originally conceived as an homage to the founder-author of general system theory, includes contributions in all the major disciplines in which general system theory is the ideological frame of reference. In addition, this book contains a complete bibliography of Von Bertalanffy's scientific writings, including papers and articles.

Rizzo, N. D. The significance of von Bertalanffy for psychology. In E. Laszlo (ed.), *The Relevance of General Systems Theory.* New York: George Braziller, 1972(a), pp. 135-144. All the papers in this volume were presented at an interdisciplinary symposium, held at the State University of New York, Geneseo, on September 18, 1971, honoring Ludwig von Bertalanffy on his seventieth birthday.

––––––. Theory and results of non-voluntary psychotherapy. *International Journal of Offender Therapy and Comparative Criminology*, 1972(b), *16*(2), 154-159. Presented in Italian as "Il principio fondamentale della psicoterapia non-volontaria," at the International Congress of Psychotherapy, Milan, August, 1970.

––––––. The court clinic and community mental health: General system theory in action. In W. Gray and N. D. Rizzo (eds.), *Unity Through Diversity.* New York: Gordon & Breach, 1973, Vol. 2, pp. 897-925. Presented in John Umstead Series of Distinguished Lectures, Raleigh, North Carolina, February, 1970.

Rizzo, N. D., Gray, W., and Kaiser, J. S. A general systems approach to problems in growth and development. In W. Gray, F. J. Duhl, and N. D. Rizzo (eds.), *General Systems Theory and Psychiatry.* Boston: Little, Brown, 1969, pp. 285-295.

Sheldon, A., Baker, F. and McLaughlin, C. P. (eds.). *Systems and Medical Care.* Cambridge, Mass.: M.I.T. Press, 1970. The published papers were presented at the Symposium on Systems and Medical Care held at Harvard University in 1968. The selection of papers is excellent, and the book can be recommended highly.

Wheeler, R. H. *The Laws of Human Nature*. New York: Century, 1932.

CHAPTER **3**

The Client as the Focal Subsystem

Gordon Hearn

Social work started out as a holistic process. In the early days of the settlements and the reformers, it was all together; social workers worked interchangeably with individuals, groups, or the whole community, as the situation demanded. But then they began to specialize — first in their work with individuals, then with groups, and finally with neighborhoods and the community. Social work began to be taught and practiced as three methods — casework, group work, and community organization. Professionals specialized in one or another and, for a time, at least, tried not to cross the boundary of their specialties. Thus, social work was no longer holistic but was divided into several parts.

This specialized approach has proved inefficient; it is not the way the client lives his life, nor is it the most effective way to foster improvement of the human condition. Consequently, in recent years there has been a major trend toward putting social work back together. We are striving to devise a fundamental theory of social work intervention, one that would serve to guide us in all social work situations, at any particular moment of intervention, whether at the level of the individual, the family, the peer group, the organization, the neighborhood, or the larger community. Many professional schools are now trying to produce generalists.

My view has been that our success in developing such a basic conception of the social work process depends upon our ability to develop an equally basic and universally applicable conception of the client system. We need a model of the client that will be equally serviceable whether the focal client is an individual, a dyad, a triad, a small group, or some larger social system.

Some years ago I worked on such a task with a group of students at the School of Social Welfare, University of California at Berkeley (Bolter et al., 1962). This group developed two models—one of the system universe, the other of the human system with its internal resources and their relationships. Both models can be used to represent the client as the focal subsystem.

The concept *subsystem* should be clarified at the outset. The term is used in two ways in general systems theory: Sometimes it is used to refer to a structural or functional process in a system — for example, the circulatory system in an individual or the communication system in a group. But it is sometimes also used to refer to one of the components of which the system is composed, such as one of the members of the group. Young (1964) used the term *special purpose subsystem* to refer to the first usage and *general purpose subsystem* for the second. I will use subsystem in the latter sense, as a general purpose subsystem.

Any general purpose subsystem can also be regarded as a system in its own right. Thus, when I refer to the client, I view him both as a system and as the focal subsystem in a larger system. The two Berkeley models are of systems, but they can also refer to the client as the focal subsystem.

THE SYSTEM UNIVERSE

The model of the system universe is represented in Figure 1. The system is the area within the triangle and has two subdivisions: the area outside the circle — the part of which the system is aware — and the area inside the circle — the part that is unknown to the system. Similarly, there are two regions in the environment: the distal environment outside the square and the proximal environment inside the square. The proximal environment may be defined as the environment of which the system is aware, whereas the distal environment affects the behavior of the system but is beyond the system's awareness. The model is appealing because it applies equally well to all levels of the human system to which it is referred — individual, group, organization, or community.

The Berkeley group drew heavily from *The Dynamics of Planned Change* (Lippitt, Watson, and Westley, 1958), which follows a general systems approach. This book, which concerns planned change with efforts directed toward individuals, groups, organizations, and communities, proposes a most useful typology of systemic problems. According to the authors, a system, in order to maintain viability, copes with problems, some of which are internal to the systems and others which are external. The internal problems are (1) the distribution of energy; (2) the mobilization of energy; and (3) communication within the system. The external problems are (1) achieving correspondence between internal and external reality; (2) setting goals and values for action; and (3) developing skills and strategies for action.

Figure 1. The system universe.

Those in the health and welfare professions assist their clients by helping them deal with any and all of these systems problems. The Berkeley group was particularly interested in the first of the external problems, that of achieving correspondence between internal and external reality. As they thought about their work as professional change agents in relation to their model of the system universe, they concluded that the Lippitt, Watson, and Westley formulation might be changed so as to achieve correspondence between the *pairs of regions* in the system universe. Because there are four regions, there are the following six pairs of regions: (1) conscious and unconscious regions of the system (O and △); (2) the conscious region of the system and the proximal environment (▽ and □); (3) the proximal and the distal regions of the environment (□ and ☼); (4) the unconscious region of the system and the proximal environment (O and □); (5) the unconscious region of the system and the distal environment (O and ☼); and (6) the conscious part of the system and the distal environment (▽ and ☼). Their attention was drawn also

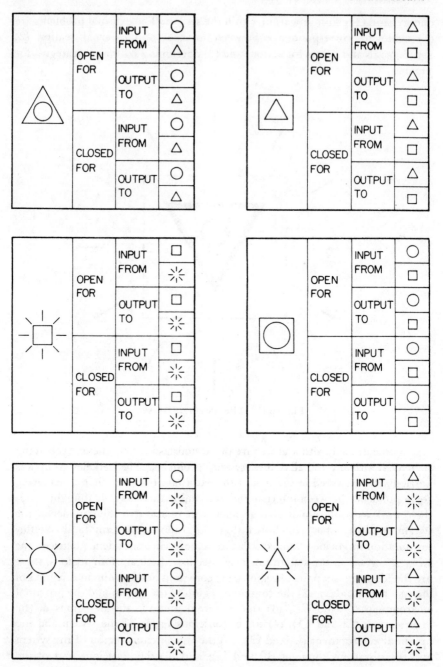

Figure 2.   Boundary relationships between regions of the system universe.

to the boundary between any pair of regions, particularly to the degree of openness and closedness for input from one region to another and output to one region from another. These boundary relationships between regions of the system universe are summarized in Figure 2.

The openness or closedness of the boundary between any two regions is controlled from both sides, as is the case of a door between adjoining hotel rooms that can be locked or unlocked on either side. Only when there is correspondence between conditions of the boundary will it be fully closed or fully open. Openness on one side and closedness on the other describes a condition of conflict, of lack of correspondence between the two regions with reference to boundary conditions. Therefore, it can be said, in reference to the first block in Figure 2, that the conscious region of the system $\triangle$ is closed for input from the unconscious $\bigcirc$, but the unconscious $\bigcirc$ is open for input to the conscious $\triangledown$. Use of the scheme, then, is in part defined by whichever side of the boundary is being described. It is meaningless to read from the scheme, for instance, that the conscious part of the system is open for input from itself or is closed for output to itself, unless such statements are useful representations of feedback processes that are internal to the region.

THE HUMAN SYSTEM

In the second model, the human system, the Berkeley group attempted to develop a representation of the human system that would serve at all levels of human organization (Figure 3).

This model was based on some earlier work I had done in trying to identify the universal properties of groups. I had postulated that there were certain phenomena that would exist in one form or another and in some state of development in every human group. These phenomena would be regarded as special-purpose subsystems. There is always an energy subsystem and a value subsystem to account for the directed action of the group. There is always a structure in the form of interrelated positional, role, and relationship subsystems, and there is always a functional aspect of the group in the form of interrelated communication, control, and locomotion subsystems. Finally, there is always a maintenance subsystem that strives to keep the system in functioning order.

The Berkeley group, using the general systems approach, reasoned that if these were truly universal properties in groups one ought to be able to find their counterparts in individuals, organizations, and communities. An explanation of their formulation, shown in Table 1, is included in their project report (Bolter et al., 1962). It is very similar in intent to Miller's attempt (1965a,1965b) to develop a universal model of all living systems.

TESTING THE MODELS

The Berkeley group's models were tested in two ways. The first was to determine whether the diagnostic typologies commonly used in social work practice could be readily translated into the language of concepts suggested by the two models. The second was to determine whether the models could be used equally well to describe what might be dysfunctional in the client systems with which we work when these systems represent several levels of human organization.

At the time the study was done, social workers used several diagnostic typologies representing several frameworks to classify problems commonly encountered in practice. These included three functions of a social work model developed by Boehm (1958); a typology within a psychiatric framework by Cockerill, Lehrman, Sacks, and Stramm (1956); two schemes for classifying family problems, one by Voiland and Buell (1961) and the other by another research group at Berkeley under the direction of Scott Briar (Aguirre et al., 1960); and a fomulation by Ripple and Alexander (1956) that categorized problems as primarily social, social psychological, or psychological.

A few examples will illustrate how some commonly used formulations can be translated into the terms suggested by the two Berkeley models outlined earlier. One of the functions of social work that Boehm identified was restoration. Translated into general systems terms, restoration can be seen essentially as regular input and output within the system, its known part, the proximal or distal environment, or any combination of these three elements.

In Cockerill's scheme, a treatment goal for a normal individual facing severe external socioeconomic problems would be the provision of social resources. An example might be a normal individual with little education and training caught in an unemployment cycle. If we acknowledge the obvious relationship between provision of social resources and input and output, we can see this example essentially as closedness of the boundary of the proximal environment in relation to openness of the system; the system's normal output is restricted because the input from the environment has slowed down. The problem might also be viewed as closedness of the system's boundary in that its restricted input has prevented the subject's adjustment to changing conditions in the proximal environment.

In the psychoanalytic frame of reference, the concept of psychosis has often been described as the loss of ego boundaries. In general systems terms, the boundary between the system and its unknown part (∇ and $\bigcirc$) is too open for the system to function in the proximal environment ($\square$). Output from the unconscious to the system, if too great in quantity, would result in input overload on the system, and the entire system as a functioning organism in relation to its environment would break down. This would explain the

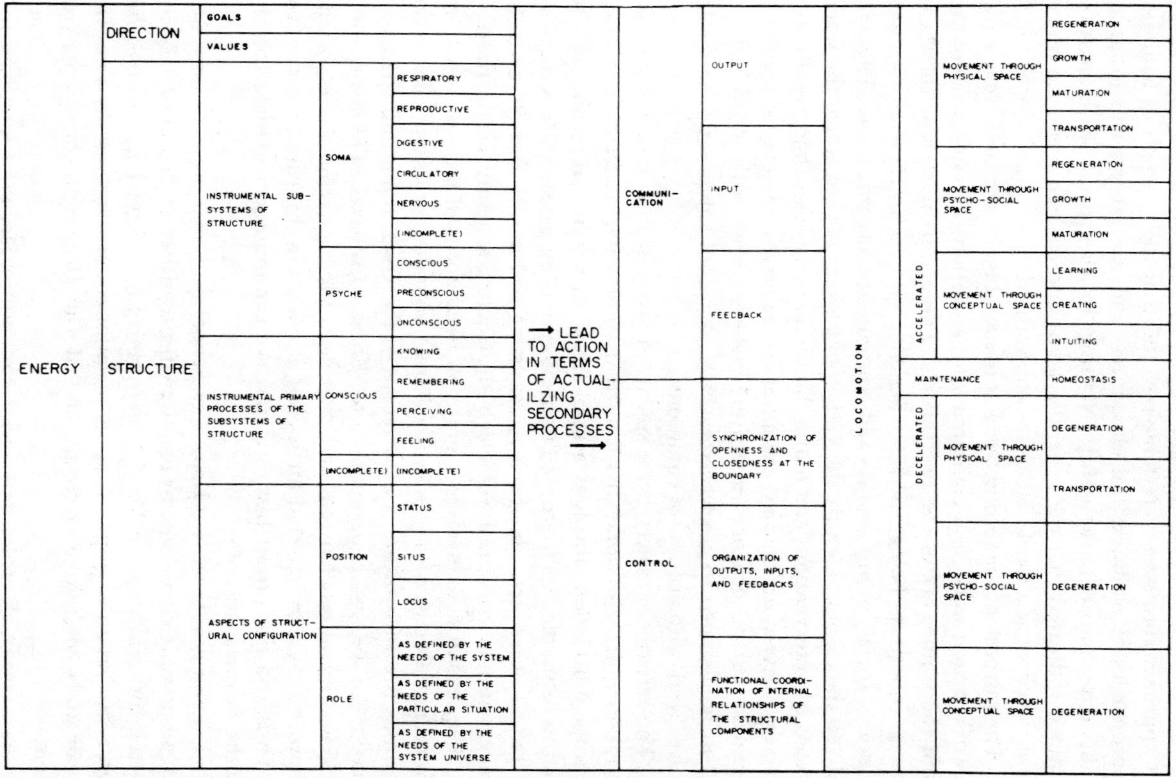

Figure 3. The human system, its internal resources and their relationships.

DIRECTION				
ENERGY	GOALS			
STRUCTURE	VALUES			

INSTRUMENTAL SUBSYSTEMS OF STRUCTURE
- SOMA: RESPIRATORY, REPRODUCTIVE, DIGESTIVE, CIRCULATORY, NERVOUS, (INCOMPLETE)
- PSYCHE: CONSCIOUS, PRECONSCIOUS, UNCONSCIOUS

INSTRUMENTAL PRIMARY PROCESSES OF THE SUBSYSTEMS OF STRUCTURE
- CONSCIOUS: KNOWING, REMEMBERING, PERCEIVING, FEELING, (INCOMPLETE)
- (INCOMPLETE): STATUS
- POSITION: SITUS, LOCUS

ASPECTS OF STRUCTURAL CONFIGURATION
- ROLE: AS DEFINED BY THE NEEDS OF THE SYSTEM, AS DEFINED BY THE NEEDS OF THE PARTICULAR SITUATION, AS DEFINED BY THE NEEDS OF THE SYSTEM UNIVERSE

↑ LEAD TO ACTION IN TERMS OF ACTUALIZING SECONDARY PROCESSES ↑

COMMUNICATION
- OUTPUT
- INPUT
- FEEDBACK

CONTROL
- SYNCHRONIZATION OF OPENNESS AND CLOSEDNESS AT THE BOUNDARY
- ORGANIZATION OF OUTPUTS, INPUTS, AND FEEDBACKS
- FUNCTIONAL COORDINATION OF INTERNAL RELATIONSHIPS OF THE STRUCTURAL COMPONENTS

LOCOMOTION

ACCELERATED
- MOVEMENT THROUGH PHYSICAL SPACE: REGENERATION, GROWTH, MATURATION, TRANSPORTATION
- MOVEMENT THROUGH PSYCHO-SOCIAL SPACE: REGENERATION, GROWTH, MATURATION
- MOVEMENT THROUGH CONCEPTUAL SPACE: LEARNING, CREATING, INTUITING

MAINTENANCE: HOMEOSTASIS

DECELERATED
- MOVEMENT THROUGH PHYSICAL SPACE: DEGENERATION, TRANSPORTATION
- MOVEMENT THROUGH PSYCHO-SOCIAL SPACE: DEGENERATION
- MOVEMENT THROUGH CONCEPTUAL SPACE: DEGENERATION

behavior of the schizophrenic and the "psychotic group" as, for example, the lynch mob in which the collective unconscious aggressive impulses are apparently operating. Similarly, a "psychotic community" would be explained in similar terms.

The Berkeley research group concluded that it was possible to translate each approach into systems terminology and that the concepts of boundary, the properties of openness and closedness, and the processes of input and output were particularly useful. Nothing in the social work literature examined indicated that the general systems approach to a generic conception of the client as a focal system was impossible.

The purpose of the second test was to see whether one could take typical casework, group work, and community-organization cases and analyze them within the framework of the two models. The researchers chose three cases from a collection being used at Berkeley to teach, respectively, social casework, social group work, and community organization. They first analyzed the cases to identify the variety of systems toward which the worker directed his interventions and found that in all three cases there were systems at several levels of human organization. This fact alone supports the argument that no practice exclusively involves "work with individuals," as used to be assumed in casework teaching, nor does any practice exclusively involve "work with groups" or communities.

The researchers selected for special analysis in each case three or four systems in which the worker had noted dysfunctional patterns. The first, a casework illustration, involved a man from a correctional setting who was about to be paroled and returned to his family. The parolee, Mr. Cesari, was selected as one system; his family, consisting of his wife and children, was another; and the community of which Mr. Cesari was a part was the third.

The second case involved a group of neighborhood kids, the Owls, and their relationship to a long-established and outmoded community institution, known as Brown Institute. In the record of the Owls, the group changed its behavior to be more acceptable to Brown Institute. The behavior of individual members changed, and the role of the institution also changed in the process. Systems selected for special attention in this case were the Owls; the Owls and the Brown Institute; and the community, including the Owls and the Brown Institute.

The third case, entitled Kent High School, was a community study. More than 40 possible client systems could be distinguished in this record. Systems selected for analysis were Mr. P., the principal of Kent High School; Kent High School as an example of an organizational-client system; the Kent High School study committee, a group; and Morlin District, a community-client system.

As is true in all professions, social work utilizes jargon in describing social process. The Berkeley group attempted to talk about the focal client systems selected from these cases in general systems terminology and in relation to the group's two models.

Their report summarizes each record and demonstrates the applicability of their models. The first case, Mr. Cesari, the parolee, can be translated into systems terminology as follows:

Mr. Cesari as a client system can best be described in terms of the model of the human system. Stated first in ordinary terms, his behavior has been described as "noisy and childish." He appears to be verbally aggressive, very manipulative, and in need of a clearly defined structure in which to operate. He is unreliable, full of ambitious, unrealistic plans, and manipulates chiefly through lying. He does not learn readily through experience and acts out repetitively.

Considering the system universe, Mr. Cesari, as a human system, is too closed for input from the proximal region in terms of the conditions of his parole. He malingers on his first job, goes further into debt, purchases a car, and does not keep the parole officer informed of his actions.

In terms of the human system, note can be taken of an apparent lack of correspondence between conscious and unconscious goals. Unconscious goals lead to repetitive actions which are part of the human system and with what appears to be the expectations of the proximal region of the system universe. Mr. Cesari's actions in relation to the system universe are determined by these unconscious needs. There is dysfunction in the structural subsystems, manifesting itself in problems of internal control (where the ego is unable to exert control over unconscious impulses). There are problems of internal direction (where the ego and superego have not been able to direct or channel expression of unconscious impulses. Locomotion through psychosocial space is decelerated (Bolter, et al., 1962, pp. 72-73).

In summary, I will suggest the potentials and the limitations I see in this undertaking.

First — the accomplishments and potentials for future work:

1.   The two models, when used in combination, constitute the framework of a generic language with which the structure, the functioning, and the dysfunctioning of human systems at all levels of human organization can be described.

2.   The model of the system universe relates the system to its environment and makes possible the expression of a whole range of influences that are felt by the system and affect its functioning but are not internal to it.

3.   Models can be used interchangeably among the several client systems in a given situation and among the various levels of human organization represented in the several client systems selected for attention.

4.   The models, when more fully refined, provide a basis for a generic conception of professional intervention.

5.   These models, conceived in relation to social work, could be just as readily used in any of the medical or social helping services. Perhaps they pro-

vide a basis for a truly general and interprofessional conception of intervention.

6. The models provide a basis for defining health and pathology. In relation to these models, health, in part at least, can be viewed as the extent to which the system has achieved correspondence among the various regions of its system universe.

7. The models may also provide a basis for elaborating the nature of the creative process. Creation is possibly the process of bringing into the known what has hitherto been in the unknown.

8. The models focus attention upon the boundary between the various regions of the system universe and the various subsystems. This suggests that much of professional intervention might be regarded as boundary work.

Second — I must also acknowledge some limitations of the models:

1. They are still in a primitive state of development and as such are currently useful for explanation, but they are probably not useful for the generation of theoretical propositions.

2. They provide a guide to professional judgment but are not a substitute for it.

3. In their present state they probably lead to few, if any, new insights by professionals. At best, they provide a useful way of ordering what is already known.

4. And, finally, they do not provide a complete analogy to all aspects of all human systems. The primary difference lies between the models and the several levels of human organization. The model of the human system that was developed from a formulation used primarily to analyze groups is clearly not as applicable at the individual, organizational, and community levels. The previously cited work of James G. Miller (1965a, 1965b) goes much further toward this end.

REFERENCES

Aguirre, A., Hanlon, D., Johnston, D., Lohmus, N., Miller, M., Spencer, M., Stevens, C., and Turner, B. Toward a diagnostic typology for the family. Unpublished group research project, University of California at Berkeley, 1960.

Boehm, W. W. The nature of social work. *Social Work*, 1958, *3*(2), 10-18.

Bolter, A., Carter, A., Hill, G., Hinckley, D., Jansen, R., Kusumoto, S., Myers, J., Philbrick, B., Rouse, L., Schmalie, G., and West, J. Toward a generic conceptualization of human systems. Unpublished group research project, University of California at Berkeley, School of Social Welfare, 1962.

Cockerill, E., Lehrman, L. S., Sacks, P., and Stramm, I. *A Conceptual Framework for Social Casework*. Pittsburgh: University of Pittsburgh Press, 1956.

Lippitt, R., Watson, J., and Westley, B. *The Dynamics of Planned Change*. New York: Harcourt, Brace, 1958.

Miller, J. G. Living systems: Basic concepts. *Behavioral Science*, 1965, *10*, 193-237. (a)

———. Living systems: Structure and process. *Behavioral Science*, 1965, *10*, 337-379. (b)

Ripple, L., and Alexander, E. Motivation, capacity, and opportunity as related to the use of casework service: Nature of client's problem. *Social Service review*, 1956, *30*, 38-54.

Voiland, A. E., and Buell, B. A classification of disordered family types. *Social Work*, 1961, *6*(4), 3-11.

Young, O. R. A survey of general systems theory. *General systems: Yearbook of the society of general systems*, 1964, *9*, 61-80.

PART TWO

Health Professionals as Subsystems

The papers by Tenney, King, and Hearn describe the functions of three health professionals—the nurse, the physician, and the social worker—as intervention subsystems within the health delivery system.

King presents her conceptualization of nursing within the framework of general systems theory. She discusses three levels of nursing function: the individual, the group, and society. She then describes these levels in terms of interacting systems—the individual as a personal system, the group as an interpersonal system, and society as a social system—and analyzes the nurses' expectations within each of the three systems. King emphasizes the importance for nursing, as well as for other health fields, of utilizing health models rather than disease models in the conceptualization of health care. She presents five examples of nursing situations to emphasize the need for research in nursing and health care to support the use of systems theory and to demonstrate her belief that it is essential to go the systems route to develop theories and conduct research in human behavior.

Tenney's paper introduces ideas for health service research. The author discusses the medical intervention subsystem and

37

offers his interpretation of the physicians' role, the objectives of medical intervention, the characteristics of the environment in which the physician functions, resources employed by the physicians to perform their functions, the individual physician as the elemental component of the subsystem, and controlling factors within the subsystems.

Medical intervention theoretically has dual functions: direct provision of personal health care service and indirect provision of informational and management support services. Tenney identifies the direct provision of personal health care as the primary clinical task, problem processing within the doctor-patient relationship. The support task includes management or information processing directed toward patient care and also toward community health maintenance, health education, and research.

In discussing boundaries imposed on the medical intervention subsystem, Tenney emphasizes that the expectations of patients and society in general significantly influence the performance of the subsystem. He also notes the nonspecific boundaries of intervention set up by accommodations among allied health professionals.

Primary care, secondary care, and tertiary care are the three principal levels of the health service system organization through which the activities and clinical task of the medical intervention subsystem can be studied. Tenney describes the activities of physicians within the three levels and identifies problems in the organization and delivery of service. He comments on the controlling factors within the medical intervention subsystem and urges the use of a systems approach to develop acceptable and effective quality control techniques in providing quality service. He comments on the controlling factors within the medical intervention subsystem and urges the use of a systems approach to develop acceptable and effective quality control techniques in providing quality service.

Hearn presents three holistic conceptions of social work— as boundary work, as human recycling, and as maintenance work. William E. Gordon's work is the basis of Hearn's discussion of social work as boundary work. Hearn shares Gordon's seven basic ideas and concludes that every feature of Gordon's formulation "suggests that social work occurs at the boundary between the system and its environment and that, in this sense, social work is boundary work." Hearn then discusses the kind of involvement that occurs at the different boundaries.

Social workers help human systems locate and define their boundaries, regulate the degree of openness and closedness of the boundary, regulate the content and rate of what enters and leaves the system, regulate the form in which matters and ideas leave and enter the system and, finally, social work helps to determine "how much is to be included within their boundary."

Hearn deals with the second conception, social work as human recycling, in relation to his efforts to understand our current welfare system and the ways in which it can be improved. He presents a three-stage model that represents the process of human recycling. His second model demonstrates the interaction of several human systems that each give, seek, and receive.

Social work as maintenance work suggests that it is a support function. Hearn describes this concept using the maintenance categories of boundary, resource, systems, component, feedback, and external relations maintenance. He proposes that within any group or organization there are two kinds of leadership, task and maintenance, and that the social worker is the appropriate specialist to assume the maintenance function.

CHAPTER 4

The Medical Intervention Subsystem and Health Services Research

James B. Tenney

The health services system in this country is a large and complex disorganization. It is condoned by an ambivalent society that cannot do without the services it purveys, as desirable as doing so might seem. Nor are there good prospects in the offing to do away with the untimely death, disease, disability, or even discomfort to which man is susceptible. The logical alternative for society, then, is to reduce their risk of occurrence by maintaining the most effective countermeasures possible, without sacrificing maximal satisfaction, happiness, and other socially desirable values. By this mixture of ambivalence and logic, a diffuse health care system has been differentiated and delegated to provide necessary and desirable services for promoting health, preventing and treating disease or disability, and alleviating discomfort at all levels, consistent with an allocated share of scarce resources. The systems approach helps to characterize various system components, to understand subsystem functions and relations, and to raise researchable questions facilitating management, planning, and accountability.

I will describe one unit of the health services system, the medical intervention subsystem, by following the systems approach advocated more generally by Churchman (1968); in the process I will introduce selected areas of interest for application of research in health services. I hope that the technique will appear practical for suggesting important problems and applying scientific methods to their solution; incomplete or missing information is the major impediment to a more systematic systems approach.

The medical intervention subsystem can be defined as the professional

41

function of physicians, or what doctors do in performing their role. In general systems terms, this means that it is a process unit as distinguished from a component or structural unit of the health services system, and it serves the purposes of a more comprehensive system in special and perhaps mysterious ways that are identifiably different from the contributions of other process units, such as nursing and social work. As subsystem, it is an integral part of the larger system, but at the same time it partakes of the qualities of a system itself and can be considered as "a set of units or elements that are actively interrelated and that operate in some sense as a bounded unit" (Baker, 1970).

Conceptually, medical intervention is considered to be a dual function, consisting of the direct provision of personal health care services and the indirect provision of information and management support services. This idea reflects a medical propensity to intervene in affairs that may not relate to patient care but should be considered nevertheless.

The first, or clinical, task is patient care or problem processing. Its elemental unit is the doctor–patient relationship, a transactional social system model initially described by Henderson (1935), elaborated by Parsons (1951), and by Szasz and Hollender (1956), and more recently discussed by Bloom (1963). Basically, the doctor–patient relationship consists of a two-way transaction of pro forma and personal expectations, with psychosocial determinants including both parties' personalities, roles, reference groups, and subcultures in a sociocultural environment matrix. For the patient, antecedents are the multiple factors related to perception of a need for medical intervention, translating it into demand and assuming the patient role; for the physician, they are the acquisition and maintenance of knowledge and skills to analyze and manage patient problems effectively.

The second, or support, task is management or information processing. It is an essential part of medical intervention directed toward patient care but also toward total community health maintenance, health professional education, biomedical and health services research, and social policy direction. Moore (1970) particularly emphasized information support and described the entire medical care system as a predominantly informational one, depending "largely on the acquisition, storage and interpretation of information by both the patient and the doctor" (p. 158). His model is appealing in its simplicity and appears useful for a wide class of patient care situations. However, the entire medical intervention function must also include noninformational professional activities (such as surgery and "the laying on of hands"), as well as other informational professional activities less directly related to patient care.

OBJECTIVES

To determine what the purpose of the medical intervention subsystem is, it is necessary to specify objectives for the physicians' professional function in operational terms. This is a difficult requirement for physicians, just as it is for any professional, but particularly so because doctors are not accustomed to thinking in quite the same terms. Physicians as individuals hold both implicit and explicit objectives, as do physicians as a professional group, and all of these should be recognized. At the individual level, doctors in the main explicitly aim to do good; "to cure sometimes, to relieve often, to comfort always," is a folk saying from the fifteenth century that expresses this aim concisely. Implicitly, like others in society, each also aims to do well, or to satisfy his own personal, psychological, and social demands in the context of his own environment. At the collective or professional group level, explicit objectives are to provide personal medical services needed by individuals in society, as well as contribute to the store of medical knowledge and transmit it to students. Implicit group objectives are the preservation of professional autonomy, power, and prestige to the extent that this is possible within the profession's social environment. To perform the medical intervention function operationally, the majority of doctors render their professional services in direct patient care, and to a lesser extent (frequently by delegation to specialists among them), they conduct research, administrate, and teach. By doing so, the objectives of the subsystem are served, as well as the objectives of the entire health services system, which are even more difficult to express without resorting to phrases such as those of the WHO definition for health.

ENVIRONMENT

The medical intervention subsystem operates in an environment that imposes boundaries and constraints on its activities. The boundaries and constraints include relationships with patients, with the community or society, with the level of professional knowledge and technology, and with other subsystems within the system, as well as with certain conventions and rules for procedure.

The expectations of patients and of society, in general, have a very profound boundary influence on performance by the subsystem. L. J. Henderson is quoted as saying, "Somewhere between 1910 and 1912 in this country, a random patient, with a random disease, consulting a doctor chosen at random had for the first time in the history of mankind, a better than fifty-fifty chance of profiting from the encounter" (Blumgart, 1961, p. 449). Odds probably have improved since then, for the advent and subsequent publicity of achievements in medical science have been accompanied by rising expectations on the part of the public in terms of demands for professional services per unit of population. This level of expectation is an interface between physicians and society that many professionals would prefer to reduce toward a level of professionally-defined need, instead of the level of public demand.

Another boundary affecting patients and physicians is imposed by the level of existing knowledge and technology. A patient's knowledge about himself, what he considers normal, and what he can expect of the doctor — as well as a doctor's knowledge about the patient and about technological developments that are available and applicable to his case — affect the interchange between them. (For example the patient, a lady, says "Doctor, I hope you can treat what I've got!" and the doctor replies "Lady, I hope you've got what I can treat!") Thus faith in the physician and faith in his treatment are important ingredients of success for the placebo effect in therapy. But in addition to his actual knowledge at any one time, the existence and potential availability of a body of appropriate knowledge and technology on which the doctor might draw also presents a continually expanding boundary for medical action.

The role of other intervention subsystems interrelated to the medical intervention subsystem presents a boundary more nearly like a zone than a definitive line. Medical intervention has been defined for present purposes as what doctors do, but in any specific context the doctors' particular activities might well be done by someone else, or another subsystem's function might be assumed by the physician. In this sense the intervention process is nonspecific. The health care suprasystem is flexible enough to accommodate any number of allied health professions or lay professionals who may undertake different aspects of patient care and health maintenance in suitable situations. The nature of appropriate education and training, accreditation, and evaluation remain to be determined. Whether the question is one of substitution or transference is probably less important than whether flexibility in relations between subsystems and a clear understanding of their objectives are maintained.

Further constraints and boundaries are quite real but more difficult to depict. Effects of organizational features in medical practice, such as facilities and institutional arrangements, methods of patient payment and physician remuneration, communication and coordination between and within subsystems, public accountability and record keeping, are the objects of both conjecture and research at all the different levels of the health care system. Legal considerations involving certification and licensing, incorporation, and malpractice need to be examined. Also, a number of procedural rules stemming from tradition or conventional wisdom remain and should be studied. For instance, why should medical school last four, not three or five years, and the curriculum be essentially the same for all students? Why should hospital house staff receive low wages for long hours of work? Why should obstetricians perform normal deliveries and pediatricians provide well-baby care? A particularly important constraining procedural rule, at least in situations where resources are limited or scarce, is the widespread belief that in all cases each individual patient should receive the best known care in full measure, when in fact doing what is best for any individual may not be doing what is best for the subsystem, the system, or the society they serve (Rhodes, 1970).

RESOURCES

The resources employed by the medical intervention subsystem to perform its function consist generally of varying mixes of land, labor, and capital in the forms of technology and skills as well as equipment, facilities, and supplies. While it is almost impossible to assign a dollar value to the subsystem's function, we do know that in fiscal 1970 health expenditures were 7 percent of the American gross national product; about one-fifth of this amount, or $12.9 billion ($62.37 per capita), was for physicians' services alone (Cooper and McGee, 1970). Expenditures for other health services, supplies, research, and construction are excluded from the figure, but some proportion of spending for these should also be allocated to the medical intervention subsystem. About one third of the nearly 320,000 active physicians in the country are not included as providers of direct physicians' services, being employed by institutions and government or engaged in teaching, administration, or research.

Information is a major resource for medical intervention and probably is the basis for most of the clinical and supportive professional activities of physicians. In direct patient care, information about patients' problems in their total context is essential for diagnostic, therapeutic, and management decisions. In population health care, information about the whole community, and the epidemiologic picture of biological, sociocultural, and economic determinants of disease incidence and prevalence or of medical care utilization are necessary. Evaluation of the medical intervention function requires information about characteristics of the process, its efficacy, effectiveness, efficiency, and outcome. The store of medical knowledge and technology is drawn from and added to with each patient–physician contact. Biomedical research continually adds results, and health services research has contributed usefully. Nevertheless, information remains an underdeveloped resource, and the need for additional technology to optimize information availability, usefulness, and application within the health care system has been pointed out by clinicians and researchers alike (Bunker, 1970; White, 1967).

Another resource should be noted in passing: goodwill, or public trust in the medical profession. There is substantial evidence of increasing dissatisfaction with the medical intervention function for a variety of reasons, including uncontrolled cost, inequitable availability, impersonal service, and questionable quality of care. These are interpreted as system-wide problems requiring major changes for solution; contributions by the systems approach and health services research may be applied.

COMPONENTS

Individual physicians are the elemental component units of the medical intervention subsystem, just as their professional activities constitute its function. To an increasing extent, clinicians have specialized after additional education and training, with the effects of limiting the range of problems they manage, of services they provide, or of patients they treat. This balkanization of the personal health care process appears to advance physician objectives more than patient or society objectives, although some contend that quality has improved as a result. Another trend has been toward both geographical and organizational concentration. The distribution of physicians differs from the distribution of the population, particularly with respect to rural–urban and central city–suburban locations. The number of solo practitioners (who have not really been alone since the advent of the telephone and rapid transportation) is diminishing, while that of partnerships or single and multi-specialty groups is increasing (McNamara and Todd, 1970).

We can observe the clinical task of the medical intervention subsystem and study the activities and performance of the component physicians in three levels of the health services system — primary care, secondary care, and tertiary care. The medical intervention process is evident at all three levels, but integration and coordination between them are less apparent, and qualitative differences among them warrant consideration.

The primary level of patient care is practiced by generalists, internists, and pediatricians — the doctors to whom patients first turn when they have a health-related problem. Problems at this level are undifferentiated and usually common; health services research studies have shown that a little more than one-half can be assigned to any specific diagnostic label within conventional disease classifications, and a large majority (80 to 85 percent) can be managed entirely by the primary physician in his office (College of General Practitioners, 1963; Royal College of General Practitioners, 1970). Many primary physicians expend much time on physical examinations and on the (to them) trivial problems of essentially healthy patients, problems for which these physicians are not trained and in which they are not interested (Bergman, Dassel, and Wedgwood, 1966). Research efforts now underway seek to analyze the clinical tasks at this level and to determine which among them could more appropriately be delegated to other health professionals (Lewis et al, 1969; Yankauer et al, 1970). Some researchers have suggested automated multiphasic health-testing and referral service instead of the traditional fee-for-service mechanism to regulate the flow of patients at this entry point to the health services system (Garfield, 1970). Primary medical practice is the level at which the majority of physician visits or medical intervention occurs. It is, as well, the object of public and professional controversy and of interest in health services research.

The secondary level of patient care is represented in the practices of specialists, consultants, and general community hospitals—the physicians and facilities that provide services frequently mediated by referral or admission via other health professionals. Commonly, patients at this level receive special purpose care that is rendered for already defined problems of diagnosis and treatment; responsibility for continuing general care then reverts to a primary physician or to the patient himself when the special purposes have been achieved. Coordination with other levels of care, as well as the provisions for adequate supply and distribution of manpower, have contributed to problems in organizing the medical intervention function at the secondary level. Also, patients increasingly seek specialist care and insurance-covered emergency room and hospital services directly, on the one hand; on the other hand, physicians and facilities increasingly seek to extend their specialized technological capabilities at the expense of providing broad-based services. More than half the recent use of emergency rooms of a large metropolitan area was for nonurgent conditions, and an earlier national survey indicated that in more than 60 percent of hospitals with cardiac catheterization laboratories, an average of less than one such procedure was performed weekly (Webb, 1969; Crocetti, 1965).

The tertiary level of care involves the practices of superspecialists, clinical researchers, and teachers in large medical centers—the professionals and resources uniquely equipped to tackle technically complex, specialized services for difficult or esoteric problems. A large proportion of the medical intervention subsystem's support task or nonpatient care functions at this level are carried out through medical education and research. The training of future professionals for patient care by contacts with the most highly specialized interests and selected cases is justified mostly on the basis of tradition. A systems approach to the problem of curriculum design, though perhaps more useful, is rarely used, although promising changes such as those described for Massachusetts are on the horizon (Cronkhite, Alpert, and Wiener, 1971).

Physicians carry out the support task of medical intervention under different organizational frameworks. Organized medicine consists of local, regional, and national societies that are extremely conservative toward changes in the structure, relationships, functions, and financing of health care. While such attitudes may be considered dysfunctional in a sense, the efforts of medical societies and specialty organizations toward insuring licensing and certification have been helpful in establishing standards for the medical intervention subsystem. Nowadays there seems to be a trend away from the older reactionary position, and the interests of society are turning toward foundations, corporations, and health maintenance organizations to coordinate health care delivery and to promote the quality of care and professional performance. The supporting task of educating physicians and conducting biomedical research to supply new components and technology for the medical intervention function has been centered in medical schools, which constitute another organizational unit of the subsystem with even broader objectives than those of clinical education and research.

MANAGEMENT AND CONTROL

The question of who is in charge of present health care systems can best be answered by a brief review of controlling factors within the medical intervention subsystem, for there is little actual concerted management or direction in a formal sense in medicine. The individual physician exercises control over his own performance through experience with success and failure in patient care or support activities. At a level of group or peers, or other physicians with relatively equivalent qualifications, control is exerted through medical societies, hospital staffs, tissue review committees, and consultations and chart review procedures in selected group practice and prepayment plans. State licensing regulations and medical practice acts provide legal requirements commonly administered according to interpretation by societies. There seems to be little systematic control over the quality of services and care provided by the medical intervention subsystem. Current proposed legislation reflects both public and governmental concern over this problem. A systems approach to developing acceptable and effective quality control techniques would be a welcome contribution to the deliberations.

REFERENCES

Baker, F. General systems theory, research, and medical care. In A. Sheldon, F. Baker, and C. P. McLaughlin (eds.), *Systems and Medical Care.* Cambridge, Mass.: MIT Press, 1970.

Bergman, A. B., Dassel, S. W., and Wedgwood, R. J. Time-motion study of practicing pediatricians. *Pediatrics,* 1966, *38,* 254-263.

Bloom, S. W. *The Doctor and His Patient.* New York: Russell Sage Foundation, 1963.

Blumgart, H. L. Caring for the patient. *New England Journal of Medicine,* 1961, *270,* 449-456.

Bunker, J. P. Surgical manpower: A comparison of operations and surgeons in the United States and in England and Wales. *New England Journal of Medicine,* 1970, *282,* 135-144.

Churchman, C. W. *The Systems Approach.* New York: Dell, 1968.

College of General Practitioners. Prospective studies: I. Disease labels. *Journal of the College of General Practitioners,* 1963, *6,* 197-204.

Cooper, B. S., and McGee, M. National health expenditures, fiscal years 1929-1970 and calendar years 1929-1969. Note No. 25. *Research and Statistics.* Washington, D. C.: U.S. Department of Health, Education, and Welfare, Social Security Administration, Office of Research and Statistics, 1970.

Crocetti, A. F. Cardiac diagnostic and surgical facilities in the United States. *Public Health Reports,* 1965, *80,* 1035-1053.

Cronkhite, L. W., Jr., Alpert, J. J., and Wiener, D. S. A health-care system for Massachusetts. *New England Journal of Medicine,* 1971, *284,* 240-243.

Garfield, S. R. The delivery of medical care. *Scientific American,* 1970, *222*(4), 15-23.

Henderson, L. J. Physician and patient as a social system. *New England Journal of Medicine,* 1935, *212,* 819-823.

Lewis, C. E., Resnik, B. A., Schmidt, G., and Waxman, D. Activities, events and outcomes in ambulatory patient care. *New England Journal of Medicine,* 1969, *280,* 645-649.

McNamara, M. E., and Todd, C. A survey of group practice in the United States, 1969. *American Journal of Public Health,* 1970, *60,* 1303-1313.

Moore, F. J. Information technologies and health care: 1. Medical care as a system. *Archives of Internal Medicine,* 1970, *125,* 157-161.

Parsons, T. *The Social System.* New York: Free Press of Glencoe, 1951.

Rhodes, P. Conceptions of medical care: Conflict between claims of individual and of society. *Lancet,* 1970, *2*(7678), 870-872.

The Royal College of General Practitioners. *Reports from General Practice. XIII. Present State and Future Needs of General Practice.* (2nd ed.) London: Council of the Royal College of General Practitioners, 1970.

Szasz, T. S., and Hollender, M. H. A contribution to the philosophy of medicine: The basic models of the doctor-patient relationship. *Archives of Internal Medicine,* 1956, *94,* 585-592.

Webb, M. L. The emergency medical care system in a metropolitan area. Unpublished Dr. P. H. thesis, Johns Hopkins University, School of Hygiene and Public Health, 1969.

White, K. L. Improved medical care statistics and the health services system. *Public Health Reports,* 1967, *82,* 847-854.

Yankauer, A., Connelly, J. P., Andrew, P., and Feldman, J. J. The practice of nursing in pediatric offices—challenge and opportunity. *New England Journal of Medicine,* 1970, *282,* 843-847.

CHAPTER 5

The Health Care System: Nursing Intervention Subsystem

Imogene M. King

Nursing, with man as the central focus for care, is, and will continue to be, concerned with human behavior, social interaction, and social movements. An understanding of man and his environment is prerequisite to understanding the nature of nursing. As professionals, nurses deal with the behavior of individuals and groups in potentially stressful situations relative to health and illness and help people meet needs that are basic in performing activities of daily living. This may appear as a very simple statement of the role and functions of a professional nurse, but in reality when it is analyzed, it becomes very complex. One of the major reasons for nurses to develop theories and to conduct systems research is the great complexity of every nursing situation. I believe we are becoming hypnotized by the concept of change that bombards us every day. A critical analysis of nursing indicates that most of the change has been technologically and mechanically oriented rather than person oriented. Has man basically changed over the past few centuries? I think not. This belief has forced me to scrutinize nursing history, to attempt to define the nursing act, the basic element of analysis that makes up the nursing process, and then to redefine nursing in a way that I think it can be studied. But before developing this analysis, I will review the nursing profession in line with change in society.

51

CHARACTERISTICS OF NURSING OVER TIME

Transitory changes have affected nursing, but a few basic elements have remained continuously an integral part of nursing practice. Nursing continues to be a helping profession (King, 1971). Nursing provides a service to meet a social need. A part of this service is to give care to individuals and groups who are ill and who usually are hospitalized. Nurses give care to those individuals who have a chronic disease and those who need rehabilitation to help them use their potential ability to function as human beings. Nurses offer guidance for individuals and groups to help them maintain health. Nurses are partners with physicians, social workers, and allied health professionals in promoting health, in preventing disease, and in managing patient care. They cooperate with physicians, families, and others to coordinate plans of health care. Nurses are the key figures in the delivery of health services in the 1970s. Nurses are the largest health manpower group numerically and are essential in planning and providing for health care for all individuals. A part of the service provided by nurses deals with specific technical skills that they are expected to possess. Two such skills, observation and communication, are important methods for collecting information to make decisions about nursing care. Measurement of blood pressure, pulse, temperature, and respiration are gradually being programmed and information collected through monitoring devices, so that in future the actual measurement and interpretation will be made by machines, yet decisions about the use of these data will be made by nurses. The task will be abolished for nurses but the function of decision making will be retained. Skills necessary for implementing physician-initiated therapy, such as inhalation, intubation, and drainage of body cavities have been a part of nurses' repertoire for years. Some technological changes have required that nurses learn additional skills for specific situations, such as monitoring physiologic parameters of patients in specialized care units in hospitals, caring for premature infants, and giving health guidance in nurse clinics, homes, and health agencies.

Some of the technical skills involved in caring for patients in hospitals have been delegated to nonprofessionals. Thus, in seeking to identify differentiating characteristics of professional nurses, one must look to their uncommon knowledge, skills, and set of preferences that differ from those outside the group and to their legal right to perform specific functions. Many nonprofessionals can be trained to count pulse and respiratory rates and to give a hypodermic injection. Professionals, however, use the information from these techniques to plan a course of action to maintain internal and external equilibrium in patients.

One of the major changes in nursing revolves around theory development and the identification and testing of fundamental concepts that build an explicit theory of nursing. Theory development for nursing is essential if professional nurses are to continue to assume responsibilities for delivering health care services. Theoretical models can organize sets of concepts that are related and provide a rationale for gathering information that is essential for the nurse's decision making. Theories offer explanations of behaviors of individuals in particular situations and provide a framework for organization of phenomena and for studying the elements that interact with each other and are related. These reasons make it imperative for nurses and other health professionals to use systems research, because the old idea of two-variable, cause-effect type of research is not useful in dealing with current problems in the health field; neither, however, is industrial engineering the savior for systems research in nursing.

Today, health is conceived as a right for all people. This implies that health services and programs are available for individuals when they need them. Historically, nursing has provided some kind of care wherever there are people. Thus, a concept of community health nursing has always been a part of professional nursing. My theoretical framework has been derived from basic elements in nursing that have persisted over time. Man is the focus for the framework. His adaptation to life and health is influenced by internal and external environmental factors.

GENERAL SYSTEMS THEORY FOR NURSING

I am presenting a conceptual framework as a general systems theory that requires continuous testing in the real world of nursing practice. Three levels of functions include the individual, the group, and society. These levels indicate that three dynamic interacting systems are exchanging energy and information on a continuous basis. I have named these three interacting systems as follows: (a) the individual is called a personal system (or, if you will, an intrapersonal system); (b) the group is called an interpersonal system; and (c) society refers to social systems. Man, the human organism, along with his external environment, provides a central focus for these interacting systems. "Man functions in *social systems* through *interpersonal relationships* in terms of his *perceptions* which influence his life and his *health*" (King, 1971, p. 22). The personal systems of the nurse, the health client, the physician, the social worker, the physical therapist, and other related professionals are influenced by their past experiences and their knowledge, the external environmental factors of the here-and-now moments, and the goals to be achieved in the situation. Each reacts to what he is perceiving, and they interact on the basis of various sets of individual perceptions, past experiences, and the now moment of each interaction. Perception is a fundamental concept in understanding behavior. The process of knowing begins with sensory perception. Each individual is aware of events in his external environment through his own perceptions, which may be open to, selected from, or closed to external environment.

The interpersonal systems are two, three, or more individuals interacting in a given situation. A basic concept is interpersonal relations, facets of which are verbal and nonverbal communication. This situation implies a variety of perceptions in nursing situations or in health care team situations. In any interaction between two or more persons, a human process is occurring over time. I define this human process as action, reaction, interaction, and transaction. I have related this to nursing by describing the nursing act as all other human acts, that is, a sequence of behaviors of interacting persons that occur in the following three phases: recognition of presenting conditions, operations or activities related to the conditions or situations, and motivation to exert some control over the events to achieve goals (King, 1971). A series of these acts makes up the nursing process, a human process. A nursing process involves interactions of nurse and relevant others in nursing situations.

Social systems describe units of analysis in a society in which individuals form groups to carry on activities of daily living to maintain life and health and, hopefully, happiness. A hospital has been described as a social system, one established by society to care for individuals who require assistance to cope with their health.

Horwitz (1953) has suggested a three-dimensional matrix to study group behavior. He has used three systems: individual, interpersonal, and institutional. The variables one can identify in columns would be the independent variables to be studied and those identified and listed in rows would specify the dependent variables.

The interaction of my three systems — personal, interpersonal, and social — are illustrated in Figure 1. The fundamental concepts in this model shown in Figure 2 are: health as the goal for nursing, perception and interpersonal relations as two basic components of the personal, and interpersonal systems functioning in social systems.

Several questions must be faced when one looks at theoretical frames of reference for nursing or any field of study involving decision making. These include: What kind of decisions are nurses required to make in the course of their roles and responsibilities? What kind of information is essential for them to have in order to make decisions? What are the alternatives in nursing situations? What alternative courses of action do nurses have when making critical decisions about an individual's care, recovery, and health? And what tasks do nurses perform and what knowledge is essential for nurses to have in making decisions about alternatives? These questions lead to systems theory and to systems research because nursing is a complex task; because human and environmental variables appear in every nursing situation; because nur-

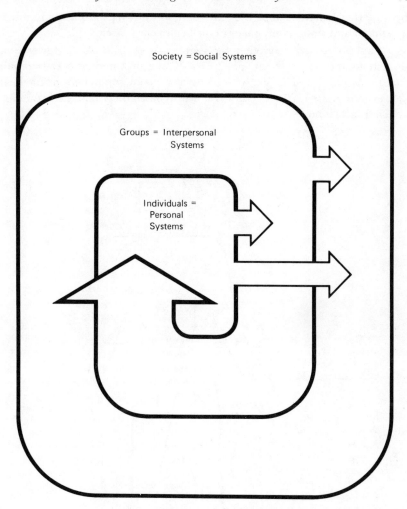

Figure 1.   Dynamic interaction systems.
(Reprinted from Imogene King, *Toward a Theory of Nursing.*
Copyright John Wiley and Sons, Inc., 1971)

ses play many roles in a variety of social organizations of different sizes and
structures, and with a variety of persons; and because the past and the present
influence the responsibilities and the decisions made by nurses. Is it any wonder then that persons in nursing, whether they are researchers, educators, administrators or practitioners, exhibit urgency and impatience? Some have a
basic knowledge of systems theory and research and seek more knowledge

because they believe it to be the only route to studying complexity in nursing as a major and viable component in the health care systems.

I am not urging everyone to do systems research, however, for emphasis on systems and especially on health care systems in American society requires that many pause and reflect on nursing as a major component in the health care system in the 1970s. The time for action based on systematic gathering of relevant facts is now.

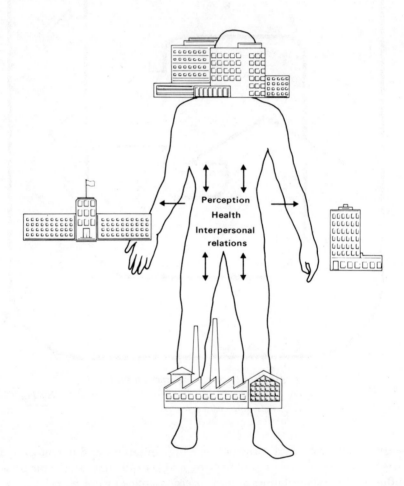

Figure 2.   Frame of Reference.
Man—health—perception—interpersonal relations—social systems.
(Reprinted from Imogene King, *Toward a Theory of Nursing.*
Copyright John Wiley and Sons, Inc., 1971.)

CONCEPTUALIZATION OF HEALTH CARE

Health models are vital if we are going to reverse the crises in health care in the United States. The illness or disease model of the past has served the very useful purpose of bringing devastating disease under control. But its usefulness as the only model is outmoded and does not help us ask the questions for today's world of rapid change through instantaneous communication. A health model sends us in a different direction and that is what a theory with a model is supposed to do, that is, tells us where to go to look for what? A health model will send us to look for factors that enhance health or impinge on it. It tells us to look for interactions between individuals, groups, and social systems. It tells us to look at normal physiology, growth, development, and performance behaviors to determine some health norms. In contrast, a disease or illness model tells us to look for and at pathology, that which is abnormal or deviant. Continued use of the medical and hospital model of sickness exclusively will deter us from asking relevant questions about health. My model for health suggests basic concepts of perception, interpersonal relations, and social systems as critical factors to study because of their dynamic interrelationships.

WHY SYSTEMS THEORY AND RESEARCH?

Five examples of nursing situations over time may help us capture the urgency for theory and research in health care and in nursing. Physicians, nurses, employers, and the public have had superhuman expectations for nurses since the turn of the century, but especially in the 1970s. Then we must consider some of the ways nurses have attempted to meet all of these expectations.

Short-term training courses have been established by a few physicians, nurses, and hospital administrators to prepare nurses to function as technical experts with machines, to monitor equipment in coronary care units in hospitals, and to learn selected diagnostic techniques needed to function as general practitioners in medicine in rural areas and remote places where there are few doctors. The technological world of hospitals demands technically oriented and technically skilled individuals to monitor the machines and the mechanical equipment, and, in most instances, nurses learn these techniques. Does an individual have to be a nurse to become the keeper of machines, and tubes, and instruments? I think not. There are levels of monitoring, however, and machines serve a very real need at certain levels. These events and other recent happenings in nursing suggest to me that it was and is expedient for physicians and hospital administrators to utilize nurses to fill a real gap in services needed by patients when they as providers of those services cannot produce the output expected by society. I have no quarrel with stopgap measures on a temporary basis that provide care for people as an immediate and emergency means. But I do quarrel with individuals who use nursing personnel (including nurses themselves) as whipping boys and take from the nursing supply to fill an emergency gap in some other field.

I think it is time for all of us in the health professions to stop reacting to every crisis and getting on every so-called innovative idea that someone dreams up. It is time to start interacting and transacting with each other to begin to get to the heart of the problem if we are going to resolve the health care crisis. What I am suggesting is that all health professionals hold the goal of individual health as the primary focus and attempt to find means to gather data to establish plans for meeting that goal in this society. It is time to look at quality and effectiveness of care; then the numbers game in manpower will not cloud the problems. We should determine what a health team of individuals can do. This is what systems theory and research is all about.

A second example of expectations we have had for nurses is as follows: We expect nurses to perform their duties in a variety of social organizations including the schools, industries, official and voluntary health agencies, and hospitals of different sizes, controls, and purposes, which employ more than 50 percent of the practicing nurses. No other group of health professionals has been asked to provide caring services in such a variety of social organizations. Within the past few years we have added nursing homes, extended care facilities, model cities health programs, neighborhood health clinics, crisis centers, drug centers, and alcohol centers.

As a third example, for at least the past 15 to 20 years nurses employed in hospitals have become distracted from their primary mission of the direct care of people and have assumed a managerial role. Within the past five years, a trend has appeared to bring the nurse back to the practice of nursing, that is, direct care of people. As the hospitals initiate changes by introducing ward clerks and unit managers to take care of administrative details so that nurses can have increased time for giving nursing care, they seldom give thought to planning for these changes and to recognizing the need of some nurses for certain kinds of retraining to enable them to return to direct care of people. Nurses have been expected to be caretakers of hospitals for 24 hours a day, seven days a week. They have been the participant observers for years and have been expected to make decisions to act on the basis of data they collected. In many instances, nurses are data collectors for physicians.

A fourth example is that over the last 50 years, the curriculums in schools of nursing have required students to take courses in various disiplines such as physiology, chemistry, anatomy, biology, sociology, psychology, and anthropology. Each discipline had a somewhat circumscribed body of knowledge (they do not seem to have such certainty now) from which faculty members in schools of nursing extracted bits and pieces of factual information. The nursing courses appeared to be organized around the medical model of categories of disease, diagnosis, and treatment plus nursing care. Students were expected to learn many facts from the variety of exposures, to synthesize these bits and pieces of information into some kind of gestalt, and then to apply the knowledge in giving direct care to people.

A fifth example of expectations is that every hour of every day, in the course of carrying out their professional functions, hundreds of thousands of nurses in this country are making observations of behavior of human beings in hospitals and in the community, Yet few of these observations are recorded in a systematic way for the preservation and study of their meaning at that moment in time or at a later time. We must discover ways to use the models that are available now (the Howland model for example), to design research protocols and secure financial support for conducting descriptive research to capture all the lost data in the health field.

These five examples of some of the expectations we have had for nurses over the past half century support the use of systems theory to develop our nursing theories and conduct our research in human behavior. In the performance of their functions in a variety of social systems, nurses assist individuals and groups in society to attain, to maintain, and to restore health. Nursing students must be introduced to new ways of ordering knowledge from the vast array of facts available to teach and for students to learn. One way is to introduce them to a theoretical framework of three systems — personal, interpersonal, and social — and to help them develop basic concepts, skills, and values. This will include a way of thinking about nursing, a systematic way of studying nursing, and ways of delivering health care through the practice of nursing. My theoretical framework is my own way of looking at complex levels of functions within three dynamic interacting systems and then defining nursing, health, and selected terms relevant to nursing (King, 1971).

Nursing in the 1970s will see a major emphasis on making explicit the role and responsibilities of professional nurses in the delivery of health care services. Trends in health care in our society will find many more professional nurses in community health nursing, that is, wherever there are people who need assistance to stay healthy. Some nurses will become highly skilled in new technologies in caring for hospitalized individuals. The role of the nurse in health teaching and in health promotion will be dramatized within the next few years. The nurse of tomorrow will represent the good of the past and the knowledge of the present, and will perform in such a way that individuals will have great hope for the future. The key to the action of living things is that it is directed toward the future.

The basic element that remains continuous throughout changes in nursing is the goal of nurses, which is health; the means nurses use to help themselves and others to achieve the goal are the assistance they offer and the care they give during interferences or crises in the life cycle. The complexity of nursing demands systems theory and research in the area.

The 1970s and 1980s present many challenges, one of which is the need to identify theories that are fundamental to nursing practice. Concepts, theories, and constructs are intellectual tools that will help nurses organize a multitude of fragmented facts into useful information. New approaches must be developed to introduce students to ways of thinking and to methods of studying, of selecting basic information, and of using knowledge. We suggest the use of systems theory and systems research in teaching students to develop fundamental concepts so that, as practitioners, their conceptual frame of reference will accommodate new knowledge throughout the lifelong process of learning. One approach within such complexity is to identify specific goals for nursing practice to determine ways that individuals and groups cope with health and illness and adapt to changes in health states.

Of necessity, any conceptualization of nursing includes knowledge and the process of applying knowledge in the practice of the profession. For a century individuals have expected nurses to integrate basic knowledge from every discipline in existence and to synthesize this knowledge into some kind of a meaningful relationship, often in a life-and-death situation, to make a decision for action, and to implement that decision.

In the years ahead nurses will have an opportunity to contribute to the science of human behavior through systems research. This cannot be achieved without collaboration with experts in systems theory and research, behavioral scientists, physicians, social workers, and allied professionals.

REFERENCES

Horwitz, M. The conceptual status of group dynamics. *Review of Educational Research*, 1953, *23*, 309-328.

King, I. M. *Toward a Theory of Nursing*. New York: Wiley, 1971.

CHAPTER 6

Social Work as an Intervention Subsystem

Gordon Hearn

Social work is a significant intervention subsystem of the larger health care system. In discussing social work in this context, I will outline three perspectives on social work—three ways of thinking about social work's special function in the larger system. Each will be an holistic conception of social work.

SOCIAL WORK AS BOUNDARY WORK

First, social work can be regarded as "boundary work" (Hearn, 1970). This view builds on the work of William E. Gordon (1969) and, in particular, on his article in a recent publication of the Council on Social Work Education. This article can be distilled into about seven basic ideas:

    1. Social work has a simultaneous dual focus. It focuses at once upon the person and his situation as well as upon the system and its environment.

    2. Social work occurs at the interface between the human system and its environment.

    3. The phenomenon that occurs at the interface is a transaction between system and environment.

    4. The transaction is a matching effort whose focus is the coping behavior of the organism on the system side and the qualities of the impinging environment on the environment side.

61

5.   Encounters between an organism and the environment leave both changed.

6.   This point is of special importance because it raises the crucial question of how we judge the outcome of an exchange. How do we know how good an outcome it is? Gordon's answer is that the best transactions are those that "promote natural growth and development of the organism and *also* are ameliorative to the environment" (p. 9), that is, making it a better place for all systems that depend upon that environment for their sustenance.

7.   The seventh point addresses itself to how this goal is achieved. Gordon suggests that entropy is the key. The answer, he says, is found in the second law of thermodynamics, which states that unattended systems proceed relentlessly toward disorder, evenness, high probability, disorganization, randomness, and continuity or what is technically called an increase in entropy.

Entropy is a constant in the universe. It cannot be destroyed; it can only be distributed differently. Thus, for growth and development to occur, there has to be a continuous redistribution of entropy between organism and environment.

On the organism side entropy has to be reduced or extracted. On the environment side the entropy extracted from the organism has to be deployed in such a way that the entropy level of the impinging environment is not itself increased — it has to be distributed in a way that is nondestructive to the environment. Energy that cannot be used by the system has to be transformed so that it is available to other systems that depend on that environment for their growth and development. This is the process that ecologists refer to as recycling. That which is useless to one organism is extracted into that system's environment in a form that is useful to other systems in the same environment. They, in turn, in using it, convert the extracted matter into a form that can be reused by the organism that originally exported it.

Every feature of this formulation suggests that social work occurs at the boundary between the system and its environment and that, in this sense, social work is boundary work. This being so, it is logical to consider the kind of work that occurs at the boundary.

When we speak thus of boundary work it is important to remember that we refer not only to work that may occur at the boundary between the system and its environment, but also at the boundary between one system and another, or at the boundary between subsystems of a system.

One of the things that social workers do is to help the system locate its boundary. It may be a matter of defining a boundary, if none is clearly perceived, or it may be a matter of reconciling the system's perception of its boundary with the way others see it, if there is a discrepancy between the two.

Social workers help the human systems with which they work to regulate their degree of openness and closedness. Systems can be so open that their integrity as a system is endangered. Or they can be so closed that they are denied sustenance in the form of materials and ideas that they need for their growth and development. There is an optimum degree of openness and closedness for each organism. Systems may vary in what that optimum degree may be. And the degree of openness that is optimal for any given moment may not be suitable for succeeding situations. Social workers help systems regulate the opening and closing of their boundaries as circumstances require.

Social workers help the human systems with which they work to regulate how much comes in and how much goes out of the system. There can be either an overload or a deficiency of materials, energy, or ideas, and this applies both to input and output. Social workers help systems develop mechanisms for regulating the flow.

Social workers help the systems with which they work to regulate what comes in and what goes out. This is essentially a filtering or censoring process. I am not suggesting that social workers should serve as censors. What they do, I think, is to facilitate the development and operation of self-filtering and self-censoring processes. I expect this is what we mean by ego development. It is the process by which the organism determines what is beneficial or harmful to either organism or the environment.

Social workers help the systems with which they work or of which they are a part to regulate the form in which matter and ideas are exported from the system to the environment or imported into it. In what form do people emerge from our welfare system or health care? Are they permanently crippled and dependent? Undoubtedly, many of them of necessity will continue to be because of the nature of their infirmity or disability. But is it not possible that many can be helped to recycle themselves, at least in spirit, into the growth-inducing processes of society? This is clearly the thrust of reeducation and of new careers for the poor. There is also a question of how the physical and social wastes of human existence can be exported into the environment. We must find ways of doing it so that instead of polluting the environment, we make available unusable material and energy for other systems that rely on that environment for their sustenance.

Social workers help the systems with which they work to determine how sharply their boundaries should be defined. This, too, is a matter of degree, depending upon the condition of the organism and of the surrounding environment. When the organism feels threatened, it is probably more natural and more functional for the boundary to be rather clearly defined, so that it is easier to determine what is in and what is out. Conversely, when the system feels secure and the environment is experienced as relatively benign, it is more likely and more desirable for the boundary to be less clearly defined. I like the boundaries of schools of social work and social agencies to be sufficiently vague so that it is relatively easy for other kindred systems to be a part of social work — to freely intermingle and collaborate. I believe that in the next decade or two social workers will find themselves in new amalgams. I like that idea, and for that reason I hope the boundaries of social work will remain vaguely defined and enticing.

Finally, social workers help the systems with which they work to determine how narrow or expansive they will be. They help to determine how much is to be included within the systems' boundaries. One way to think of growth and development is to visualize the territory within the boundary as expanding. It seems likely that there will be great variety in the rate of expansion among systems and also that the rate may vary during successive phases in the life cycle of any given system. I believe it is likely that toward the end of the system's life cycle, when positive entropy is tending to overcome the rate of production of growth-inducing negative entropy, the process is actually reversed, the boundaries may be contracted, and the territory may more likely diminish. It may be more useful for the aging system to pull in the boundaries rather than to expand them. As the energy level decreases the system may prefer being more fully a part of less to being less fully a part of more.

SOCIAL WORK AS HUMAN RECYCLING

The second way of thinking about social work is to regard it as a process of human recycling. This idea also derives from the work of William Gordon (1969), as has already been noted.

I found myself building on this idea in my efforts to understand what is wrong with our current welfare system and what would make it right. I believe that it might also help us to understand the nature of an ideal health care system.

Gordon noted that social work is a process that promotes the growth and development of the system while also being ameliorative to its environment.

Accompanying all instances of growth in a system is the accumulation of entropy or unused energy that has to be exported from the system into the environment. But this exporting has to be done in such a way and in such a form that it can be used by other systems that depend upon that same environment. Otherwise the exporting pollutes the environment and endangers the life of all systems in it. The effort should be to enhance the nurturant capacity of the environment.

Because we are talking about the welfare or the health care of human systems, we can refer to the process as human recycling.

I find it useful to think in images, and I tried to represent what social work recycling would look like in graphic form. At first I saw the welfare cycle as a two-phase process in which a client system seeks help and receives it (Figure 1).

Receiving Help

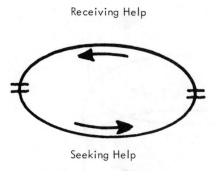

Seeking Help

Figure 1.   Two-phase process.

But then I began to consider why our welfare system and perhaps also our health care system are not working as well as they might. The two-phase cycle of seeking and receiving may become continuous; it may be entrapping, stifling, and dependency-producing.

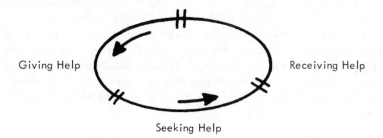

Giving Help   Receiving Help

Seeking Help

Figure 2.   Three-phase process.

The social work process can instead be thought of as a three-phase process of seeking, receiving, and giving (Figure 2). This third phase of "giving help" makes an enormous difference. It makes a giver out of the client-recipient; and it makes a recipient out of the original giver. It eases the pain of receiving and the pain of giving. Most important of all, it enhances the recipient's self-image by increasing his sense of worth. Similarly, it probably also benefits the giver by making him more humble and less condescending. Actually, what it does is turn the circle of Figure 2 into a spiral such as that shown in Figure 3.

This is not a new idea, of course. It is the essence of group therapy, of Alcoholics Anonymous and of Synanon. I believe it may be the essence of all

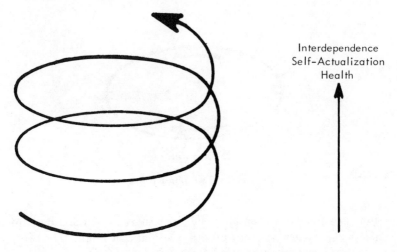

Interdependence
Self-Actualization
Health

Figure 3. The three-phase spiral.

social work and health care that *works*. I have carried this speculation one step further to produce the "Interdependency Model" (Figure 4).

Applying this model to the health care system, we can think in terms of interacting spiraling systems. One may be the focal client, another may be a doctor or a nurse, and the third may be a social worker. Each is seeking, receiving, giving, and, hopefully, spiraling upward to interdependence, self-actualization, and health. Also, the spiraling of one affects the spiraling of the others.

SOCIAL WORK AS MAINTENANCE WORK

The third way to characterize social work as an intervention subsystem is to regard social work as essentially maintenance work.

Systems require maintenance if they are to survive. It takes a ground crew, as well as an air crew, to keep an aircraft in flight. So it is with all kinds of systems—individuals, groups, organizations, and communities.

Considering social work as maintenance work suggests that it is a support function.

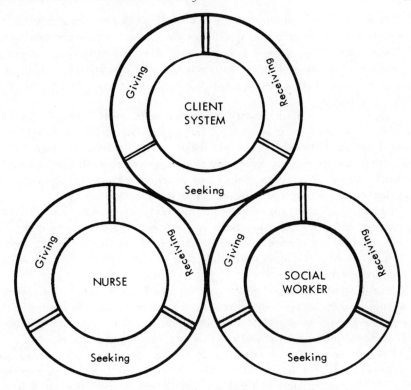

Figure 4.   The interdependency model.

In the maintenance subsystem of a larger system, four classes are being maintained: (1) the boundary of the system; (2) its resources; (3) the system itself as a system; and (4) its external relations.

Boundary maintenance involves: (1) the delineation of the boundary to indicate territoriality, what is "in" and what is "out"; (2) the regulation of inputs and outputs; and (3) the varying of the openness and closedness of the boundaries as such variation is required by changes in internal and external conditions.

Human systems as living, open, organismic systems are dependent for their survival on a ready source of people, materials, and ideas. An important function of the maintenance system, consequently, is the maintenance of resources.

While a clearly delineated and defended boundary and an adequate supply of persons, materials, and ideas are necessary, they are not in themselves sufficient for system survival. In addition, the system must be able to maintain itself in a tolerably steady state. This attempt to maintain such a steady state with all that that entails can be regarded as system maintenance.

We must first consider the nature of a system in dynamic equilibrium. It is important to realize that the equilibrium the system is striving to maintain is dynamic rather than static. A system in a steady state is not a system at rest, but rather a system in action with each of its subsystems perpetually in motion. What appears as a steady state is actually the result of a dynamic interplay of processes.

For a system to maintain itself in a steady state, every change in one of the subsystems must be accompanied by appropriate adjustive changes in many if not all of the others. If a particular set of circumstances facing the system requires a particular degree and type of control, this will require a particular structure of positions, roles, and relationships among the parts to provide the required form of communication and locomotion.

Finally, if a system is to achieve and maintain a steady state, it must have certain automatic feedback processes in operation. Whenever a change occurs in any part of the system, a signal must be transmitted through a feedback mechanism to the other parts indicating how they are to adjust in order to keep the system in balance.

Health systems require healthy components. For instance, actions taken in a group to enhance the well-being of the members will serve to promote the effectiveness of the group. Attention must also be given to maintaining the several special-purpose structural and functional subsystems.

Because feedback is also essential to keeping the system functioning as a dynamic entity, the feedback processes must also be kept in functioning order. If the wrong kind of feedback mechanism is operating, or if it is unresponsive or overresponsive, the system will be in trouble.

A human system is an open system and as such is always exchanging material and energy with its environment. As a consequence, a system must maintain a set of systemic relationships with its environment and the several relevant systems in it.

I have been saying that there are two kinds of processes at work in any living system: those that promote the system's "work" and those that maintain the system as a system.

One finds, as a consequence, that there are two kinds of leadership in a group, or an organization, or a community, or a health care system: task leadership and maintenance leadership. Further, the task and maintenance functions tend to be performed by different persons rather than by the same person. It seems natural for this kind of task and maintenance specialization to develop.

In my third formulation I am suggesting that very often and quite appropriately, social work assumes the maintenance function.

REFERENCES

Gordon, W. E. Basic constructs for an integrative and generative conception of social work. In G. Hearn (ed.), *The General Systems Approach: Contributions toward an Holistic Conception of Social Work.* New York: Council on Social Work education, 1969.

Hearn, G. Social work as boundary work. *Iowa Journal of Social Work*, 1970, 3(2), 37–42.

PART THREE

Administrative and Support Subsystems

The intervention subsystems described in the previous sections have been and will continue to be studied because of their obvious and critical relationship to patient care. However, there are many health care subsystems which receive little or nonsystematic attention because their relationship to patient care is less conspicuous. The administrative and support subsystems of health care facilities are examples of such systems.

Rosen points out that the unique tripartite organizational subsystem of many health care facilities creates unique problems and also intensifies the usual organizational problems. In such a tripartite organization decisions must often be delayed or postponed until questions of authority can be resolved. In addition, organizational conflicts may arise as the various organizational subsystems struggle for overall control of the organization. Rosen points out that occasionally the professional concerns of an individual may clash with his organizational concerns.

Rosen also discusses potential problems arising from the dual roles played by health care professionals. For example, should the health care professional also be an administrator, and, if so, to what extent should he be trained for this role? The paper outlines some of the psychological consequences of these organizational conflicts, including the problems of morale, attitude, tenure, absenteeism, and turnover.

Weisman and Mesarovic discuss an application of general systems theory to a health care problem. They develop and apply a model of a goal-seeking system to an existing problem in health care delivery—multiphasic screening—and describe systems concepts used in developing the model. They also specifically define general systems sets, functional models, systems goals, and multiphasic screening.

Weisman and Mesarovic also review the details of the screening procedure, the costs associated with the procedure, and the staffing requirements for a multiphasic laboratory. As the initial step in operationalizing the abstract model, they formulated a set of criteria by which to evaluate the performance of multiphasic screening. They then applied these criteria to selected components of the screening procedure. The results showed that multiphasic screening cannot be evaluated successfully on the sole basis of detection of asymptomatic diseases. However, since the model was developed as a dynamic system, the researchers were able to develop additional goals. They redesigned the system, using the additional goals of (1) developing a gate control for patients entering the system, (2) assisting in the diagnostic process, (3) saving physicians' time, (4) fulfilling demands for preventive health services, (5) providing tools for medical research, and (6) establishing health profiles for continuing care.

The paper concludes with a discussion of the model's ability to assist in the generation of information, which reduces uncertainty about decisions and promotes the development of appropriate goals for the system.

CHAPTER 7

The Administrative and Control Subsystems in Health Care Organizations

Hjalmar Rosen

There has been a good deal of discussion as to whether or not material relevant to administrative and support subsystems in business and industry is applicable to health care organizations. For example, Perrow (1965) pointed out that the tripartite nursing, medical, and administrative power structure of the general hospital has no parallel in the industrial organization. When we take such an observation to its logical conclusion we can see that theories, research, and general knowledge concerning the administrative subsystems of commercial organizations are not directly relevant to health care units. This point of view could be supported further by the Blau and Scott (1962) organizational typologies in which hospital organizations seem to cut across several basic categories. In addition, there is evidence that powerful substructures exist in every organization (Argyris, 1962; Cyert and March, 1963) and that to avoid organizational dysfunction generated by contradictory goals, values, and directives, the administrative subsystem must promote innovation. Coalition formation and cooptation are among the tactics that have been proposed as innovative strategies.

Although an administrative subsystem is vital to the survival and growth of any organization, the specific role and strategies of such a subsystem will vary across organizations and within organizations from time to time as a function of particular characteristics of the environment. Kast and Rosenzweig (1970) indicate that although the administration within a hospital can operate via a quasi-bureaucratic imposition of rules for many routine and repetitive functions, it must use other administrative strategies to cope with the nonroutine or unexpected. Coordination by information feedback to involved parties, linked with perceptual accuracy or role demands, plus a specified hierarchical structure (control), have been suggested as perhaps more appropriate administrative strategies under such conditions (Litterer, 1965). However, there is some evidence that, as health care units become larger in size, rules qua bureaucracy become a more prevalent administrative strategey for control than surveillance (feedback) (Burling, Lentz, and Wilson, 1956).

CONTROL: THE GOAL OF ADMINISTRATIVE SUBSYSTEMS

Perhaps the most productive way in which to discuss the managerial subsystem is in terms of its intended functions. The basic function of the administrative subsystem is to achieve coordination of the diverse organizational substructures and thereby to optimize the attainment of organizational goals and maintain the overall system via utilization of a control process.

Some persons, particularly professionals, object to the inclusion of and emphasis upon "control." Voluntary coordination, with its promised motivational byproducts and freedom from imposed directives and constraints, has considerable appeal. However, control is implicit within any administrative subsystem and within any formulation of the dynamics of organizational behavior, whether or not one finds it to be palatable. For example, control can spring from internalization generated by a forgotten reinforcement schedule (see, for example, Henry, 1949). Any efforts in either selection or training are impositions of control.

More obvious control mechanisms, however, are utilized to induce coordination. Essentially, control is initiated via a power base, i.e., resources for need gratification and/or deprivation of others. There is growing evidence, however, that the traditional power bases of reward, punishment, and legitimacy must be augmented by referent and expertise power (French and Raven, 1959). For example, Korman's (1966) review of leadership functions strongly suggests the importance of a broader power base to optimize effective organizational behavior via voluntary acceptance of control.

Obviously, however, for control to be utilized effectively in the administrative subsystem it must be augmented by adequate communication and feedback mechanisms. Moreover, insofar as control is merely an implementing mechanism, decision making is critical to establish direction and content.

DYSFUNCTIONAL ASPECTS
OF HOSPITAL ADMINISTRATIVE SUBSYSTEMS

One of the factors in effective administrative subsystem operation in health care systems which has been posited as dysfunctional is the trifurcation of the organizational power bases (Perrow, 1965). Although the trustees' function (initially the dominant power source) in the subsystem has been largely relegated to the professional administrator, the role of the medical staff in the administrative subsystem still remains as a potentially dysfunctional characteristic of that subsystem.

Some scholars have suggested that as the administrative role becomes more professional there will be a reduction in the power of the medical staff (Kast and Rosenzweig, 1970). It would seem, however, that as long as the health care system has a powerful professional class associated only tangentially with the hospital organization—in the sense of utilizing and demanding services, but without having the hospital as a professional and economic base—intraadministrative conflict among subsystems such as for resource allocation will continue to occur.

Attempts at including medical doctors as staff employees have proven interesting. (Experience with The Kaiser Health Plan in California and Community Health Association in Detroit, Michigan is supportive of this point.) Without discussing the merits and demerits of such approaches, we can note that there is some evidence that patient care cost is reduced and the scope of patient care is increased—two factors usually considered in evaluating the effectiveness of the administrative subsystem within a health care organization. We are making no plea for any particular method of reducing dysfunctional aspects within the administrative subsystem. We are suggesting, however, that as long as two sources of power and control exist within an administrative subsystem, there is the very real potential for malfunction.

The problems generated by the bifurcation of the administrative subsystem function can be discussed on many levels. Initially, however, it would perhaps be most valuable to compare parallel problems within industrial organizations. According to Argyris (1962) each major divisional head within a plant or company attemps to maximize; he attempts to influence overall policy in such a way as to maximize goal achievement of his relevant unit and pursues activity leading to his own self maximization. In contrast, organizational leadership attempts to formulate policy and pursue activity geared to optimize for the whole (Katona, 1951).

The members of the medical staff tend to use the hospital for diagnosis and treatment of their patients, as a resource of personnel and equipment no longer possible within the confines of their office or practice. However, their loyalty and concern is with their patients—not with the hospital and its functions as a totality.

On the other hand, the professional hospital administrator, although concerned with patients and their health needs, does not and cannot focus upon them as individuals. His loyalty and concern is with the overall organization. From his point of view, only through the effectiveness of the hospital as an organizational system will the focal group, the patients, optimize. He is constantly forced to make (or attempt to make) decisions that will maximize the gain for the greatest number, even if such practice may not always be optimal for a specific patient. Moreover, as is true of any organizationally-identified person, the hospital administrator often tends to put organizational welfare above welfare of either organizational constituents or organizational recipients, staff and/or patients. This is often expressed in terms of rule rigidity and generalization, which in turn may be dysfunctional for the primary goal of the organization: patient care. Unfortunately, this tendency is accentuated in a hospital setting where the administrator often is viewed as an outsider by the professional staff. The resultant insecurity leads to accentuated attempts at rigid enforcement of rules.

Tentative Solutions

One partial solution to the "built in" dysfunctional aspects of hospital administration subsystems is to develop a functional administrative hierarchy within the subsystem and to staff that hierarchy with persons trained in administrative skills. Although this has occurred in several areas of hospital administration — for instance, housekeeping and the business office — with some degree of success, much needs to be done in the administration of professional services. Although titles and positions exist, persons filling such posts are rarely trained and/or qualified to perform the administrative function effectively. By training and inclination they are health care professionals, not administrative professionals. As a consequence, not only are they unfamiliar with administrative strategies and practices, but they have an overly narrow concern for the demands of their own staffs and minimum appreciation of the problems and demands of other units within the overall system.

Possibly for optimal administrative subsystem functioning a composite administrative-medical and administrative-nursing professional training is a vital necessity. For temporary relief, however, a more feasible route would encompass in-house seminars on administrative problems, role, and techniques.

To spell out detailed areas of concentration in a brief paper would be impossible. Ideally, however, one would begin by unearthing the problem areas as perceived by organizational personnel (staff) and recipients of organizational practices (patients). The responses generated by such persons could then serve as a focal point for discussions and policy formation by the various interest groups involved in the administrative subsystem. Greater appreciation of the problems and needs of other interacting groups would be achieved through the utilization of representatives from the various organizational entities within the hospital structure in a group problem-solving approach.

ADMINISTRATIVE SUBSYSTEM FOR THE NURSING FUNCTION

Let us now turn to an example of the operation of subsystems—the administrative subsystem as it applies to the nursing function specifically.

Under the hospital administrator is an administrative subsystem for nursing personnel. Potentially dysfunctional factors occur independently of the specific organizational subunit being considered. These factors become evident whenever any health care role is brought into the large and complex organizational framework of the modern general hospital.

Historical Change in the Nursing Role

Historically, the nurse has tended to have at least one source of exterior control—the medical doctor. The nurse was responsible for carrying out the doctor's directives for his patients. The role demands emphasized patient care, and the underlying motivation was strongly characterized by a social service drive. As the demand for health services increased, due to social legislation, greater societal affluence, and the development of both public and private health insurance, two major trends in the nursing profession occurred: (1) many nurses gained greater professional autonomy in the public health field and (2) the vast majority of nursing professionals, as they became hospital employees, found themselves enmeshed in the complexities and contradictions inherent in any large organization. Although the role of the public health nurse is an intriguing one, we shall concentrate on the hospital-centered nurse.

The nurse became the employee of the hospital rather than of the doctor and therefore became subject to controls emanating from the hospital administration. But at the same time, in part due to tradition and in part to status differentials within the health care field, the doctor maintained much of his original authority over the nurse as well. Thus, the nurse was often subject to contradictory role expectations.

To cope with the administration of nursing services, new positions came into being that were professional but also administrative in function: the head nurse, the nursing supervisor, the superintendent of nursing. As was the case in early industrial development, a common fallacy prevailed in the selection and assignment of such personnel. The most effective professional tended to be promoted into administration. Not only were these nurses given little preparation for their new duties, but the new roles often did not fulfill the work needs of the promoted nurses. Some succeeded; many were inadequate; but no systematic standards had been developed for the new positions. As a consequence, efficiency of the service suffered, and dysfunctional aspects such as resource and service allocation grew more pronounced.

Administrative Aspects of the Nurses' Role

As pressures — and costs — for health services increased, less skilled semi-professionals were introduced into the patient care subsystem. To a greater and greater extent, direct patient care was assigned to and performed by practical nurses, by orderlies, and sometimes by volunteers. As a consequence, the registered nurse moved into a quasi-supervisory role and became part of the administrative subsystem — a position for which she was inappropriately selected and trained.

The administrative subsystem relating to the nursing function, then, called upon personnel to assume administrative duties for which individuals were ill-prepared and not motivated. The resultant role conflicts that ensued have been widely reported (Berkowitz and Malone, 1968; Hughes, Hughes and Deutscher, 1958; Kramer, 1968).

Solutions to the developing problems involve two critical areas of the control function: selection and training. In addition, restructuring and redefinition of the nursing function within the modern hospital is clearly a prerequisite to any formal problem-solving attempts. With the stress on ever increasing health care demands, although not necessarily optimal, modifications in selection and training are perhaps initially the most expedient source for remedial action.

Selection should stress job interest, qua motivation, not only in patient care but in supervision-administration. In part, through job choice orientation, the realities should be presented to permit sound self-selection. This should be followed by the development of appropriate and validated selection procedures. Again, as the result of realistic job analysis, training must focus upon role demands as they exist in the positions rather than on traditional and/or fallacious, romantic misconceptions of the role. This is not to say that the nurse oriented toward patient care is not a critical commodity and should not be selected and trained, but rather that the critical need for the administrator-health care professionals should be met by selection and training as well.

REFERENCES

Argyris, C. *Interpersonal Competence and Organizational Effectiveness.* Homewood, Ill.: Dorsey Press, 1962.

Berkowitz, N., and Malone, M. Intra professional conflict. *Nursing Forum*, 1968, 7(1), 51-71.

Blau, P. M., and Scott, W. R. *Formal Organizations.* San Francisco: Chandler, 1962.

Burling, T., Lentz, E. M., and Wilson, R. N. *The Give and Take in Hospitals.* New York: Putnam, 1956.

Cyert, R. M., and March, J. G. *A Behavioral Theory of the Firm.* Englewood Cliffs, N. J.: Prentice-Hall, 1963.

French, J. R. P., Jr., and Raven, B. Bases for social power. In D. Cartwright (ed.), *Studies in Social Power.* Ann Arbor: University of Michigan Press, 1959.

Henry, W. E. The business executive: Psycho-dynamics of a social role. *American Journal of Sociology*, 1949, 54, 286-291.

Hughes, E. C., Hughes, H. M., and Deutscher, I. *Twenty Thousand Nurses Tell Their Story.* Philadelphia: Lippincott, 1958.

Kast, F. E., and Rosenzweig, J. E. *Organization and Management: A Systems Approach.* New York: McGraw-Hill, 1970.

Katona, G. *Psychological Analysis of Economic Behavior.* (1st ed.) New York: McGraw-Hill, 1951.

Korman, A. Consideration, initiating structure, organizational criteria. *Personnel Psychology*, 1966, 19, 349-362.

Kramer, M. Role models, role conceptions, and role deprivation. *Nursing Research*, 1968, 17, 115-120.

Litterer, J. A. *The Analysis of Organizations.* New York: Wiley, 1965.

Perrow, C. Hospitals: Technology, structure, and goals. In J. G. March (ed.), *Handbook of Organizations.* Chicago: Rand McNally, 1965.

CHAPTER 8

An Application of General Systems Theory to Multiphasic Screening

Thomas Weisman and Mihajlo D. Mesarovic

Before we discuss the applications of systems to health research, we shall define precisely what we mean by a system. A general system, S, will be a *relation* on abstract sets:

$$S \subset V_1 \times \ldots \times V_n$$

where $\times$ denotes the Cartesian product and the sets $V_1, \ldots, V_n$ are termed system objects and denote the range of values of the attributes that describes the system. A set is merely a collection of objects—in this case, values. In an example, a system to describe the function of a light switch and the lighting fixture to which it is connected, the set V_1 contains the values which the switch can attain, on and off. Similarly, V_2 contains the values the light can attain, also on and off. This can be expressed as

$$V_1 = \{S_{on}, S_{off}\}$$
$$V_2 = \{L_{on}, L_{off}\}$$

The Cartesian product of V_1 and V_2, written $V_1 \times V_2$, is the set of pairs which can be written by taking the first element of the pair from V_1 and the second from V_2. This new set can be expressed as

$$V_1 \times V_2 = \{(S_{on}, L_{on}), (S_{on}, L_{off}), (S_{off}, L_{on}), (S_{off}, L_{off})\}$$

where S = switch and L = light.

But the system we shall call S is a subset of the Cartesian product:

$$S \subset V_1 \times V_2$$

This means that every pair in the Cartesian product is not necessarily in the system S. We therefore formulate a constructive specification for the relation between the system objects. In this example, the constructive specification is: a pair is in the system S if both the S and L have the same subscript.

$$S = \{(S_{on}, L_{on}), (S_{off}, L_{off})\}$$

At this point one might question whether what we have defined as the system S does not necessarily correspond to the situation in the real world. For example, the switch may be broken, the lamp may be burned out, or the wires may be improperly connected. Furthermore, such extrinsic considerations as a general power failure are not taken into account. However, what we have defined as a system is a *model* of some aggregate of interacting objects in the real world, not the objects themselves. A study of systems is a study of models and their properties, but these models and properties do not necessarily correspond directly to reality. In fact, it is impossible to formulate a model that takes into account all the facets of reality. In the example of the switch and the light, real-world considerations might include the possibility of a power failure that may be related to the weather, the availability of coal and oil, or the maintenance of the power-generating equipment, minor considerations, perhaps, but they do exist. And since they are minor we can eliminate them for the purpose of our model S. The art of model building (systems synthesis) is to carefully simplify the elements in such a way that usefulness and relevance are not lost.

Building systems (models) to understand the real world is selective simplification and expression of interactions among aggregated real-world objects. A study of the properties of the system we have formulated is not a study of the real world, but rather a study of our model. Deductions and inferences are drawn as a result of the structural properties of the model and must be compared to the real-world situation. In general, a system (model) is useful when the correspondence between the predictions of the system and the behavior of the real-world objects that are modeled is good.

Another useful property of systems is that one may generalize from them. Inferences made from a study of a system can be applied to any real-life situation with properties that fit the model. For example, our system S can also be used to model a water faucet and tap, a gas valve and gas flow, or an ignition switch and automobile engine. The inferences we make from studying S could apply to all of these situations. We do not contend, however, that these inferences would be equally useful or relevant in these other cases.

Many general types of systems have been described, but there is considerable difference of opinion about which models can be used in the systems analysis of real-life situations. Too often such analysis is based solely on the description of a system on an input-output, cause-effect basis. But many real-life situations do not show clear-cut input-output or cause-effect relationships. When structuring a system, one is faced with a fundamental dichotomy, that is, a model may be based on input-output or goal-seeking representations. Input-output representations are developed solely in reference to cause-effect relationships and their associated transformations. Representations of goal-seeking are developed in reference to the goals the subsystems have, or are assumed to have, and the system's function is described as working toward these stated goals. Goal-seeking descriptions may be very complicated, involving a multilayer hierarchy of subgoals. However, for quite a few complex systems this is the only type of description that fits the observations.

In the input-output approach the objects are grouped into two categories, *inputs* (stimuli)

$$X = x\{V_i : i \in I_x\}$$

and *outputs* (responses)

$$Y = x\{V_i : i \in I_y\}$$

where $\{I_x, I_y\}$ is a partition of I (I is the index set= $\{1, 2, 3, \ldots n\}$). The system is then a relation on inputs and outputs.

$$S \subset X \times Y$$

Next, a constructive procedure, a "mechanism," is provided. It is defined by means of more specific mathematical structure on the sets X and Y which associates an output (or outputs) to a given input. Our system S is an example of an input-output representation with

$$V_1 = X; V_2 = Y$$

The constructive procedure is identification of the subscripts.

Inputs and outputs again can be recognized in the goal-seeking approach, but instead of providing a mechanism to relate stimuli to responses, the system's behavior is described in terms of a goal-seeking process, i.e., as if the system responds to any given stimulus so that a given goal (or objective) is pursued. As a general example, let us assume that in addition to X and Y we postulate the following sets:

    M   the decision set
    U   the uncertainty set
    V   the value set

We further postulate the following abstract relationships between these sets:

        $P: X \times M \times U \to Y$ outcome function
        $G: M \times Y \to V_G$ performance function
        $T: U \to V_T$ tolerance function
        $R \subset V_G \times V_T$ satisfaction relation

P takes a given input, decision, and a given uncertainty into an output:

$$P(x, m, u) = y$$

G takes a given decision and output and assigns a value to the pair:

$$G(m, y) = v$$

T assigns a value to a given uncertainty:

$$T(u) = v$$

R is a relation between two values — a good example could be "less than":

$$V_1 < V_2$$

In this example a decision $\hat{m} \in M$ may be considered satisfactory if for all $u \in U$

$$G(\hat{m}, P(\hat{m}, u)) \, R \, T(u)$$

We have defined the evaluation maps and relations of G, T, and R, but not the notion of goal. G, T, and R represent an operational objective (or goal) which must be defined explicitly to be useful. It is abstractly related to the "true" goal, and it is expected that the implementation of decisions within this framework will, in reality, further the true goal. These considerations and others lead to a hierarchical structuring of goal-seeking systems that has three layers. At the first layer the subsystem searches for the satisfactory solution from the decision set M, the operational goal and uncertainties being given. At the second layer the system changes the uncertainty in the direction of letter representation with the aim of reducing the uncertainty set U. At the third layer the system translates the "true" goal into the operational goal, that is, it determines the value set V. In a complex situation all three layers operate dynamically over time.

Figure 1, a general outline of the model, shows that each layer interacts with the other two and with the process itself. As experience generates new information and as goals change, the sets V, U, and M may change. Change in any one set may lead to change in all of the others in a continuous process.

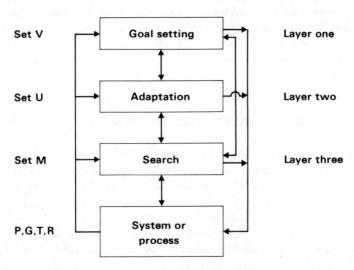

Figure 1. **Goal-seeking model.**
(Adapted from M. F. Collen, "Automated Multiphasic Screening," in C.L.E.H. Sharp and H. Keen [eds.], *Presymptomatic Detection and Early Diagnosis: A Critical Appraisal* [London: Pitman Medical Publishing Co., 1968], p. 34.)

In general, the only objects actually observed may be X and Y. M, U, and V are additional objects assumed for an appropriate and efficient specification of the system. There is little point (within systems theory as such) in arguing whether the system is actually pursuing such a goal. All that matters is that the system's functioning can be described most appropriately in such a manner. Often there is more than one way to describe the system's goal-seeking. It is also possible that the system's functioning can be described only as goal-seeking, and a specification of input-output transformation is not given. That does not mean that the system fails to satisfy some kind of causality requirements and has some intrinsically different character than, say, a system described by a set of differential equations. It only means that within the group of constructive procedures that we are currently using to describe input-output transformation, none corresponds to the observed input-output relationship. There is after all no reason to believe that all input-output transformations ought to be describable by the transformation procedures now being used. The availability of a goal-seeking description can be simply considered as a convenience, or a necessity for an efficient specification of systems (Mesarovic et al., 1970).

We shall develop a specific system model that is based on goal-seeking and that involves a real problem in health-care delivery, that of multiphasic screening. We hope to show how this model of a goal-seeking system can be used to conceptualize and explain certain events in the development of the multiphasic screening technique as a part of the health-care delivery system.

THE PERMANENTE MULTIPHASIC SCREENING MODEL

Multiphasic screening, as we shall use the term, refers specifically to a program being carried out in laboratories in San Francisco and Oakland by the Permanente (also called Kaiser) Medical Group. In the most recent year for which data are available, 47,404 patients were screened, an average of about 500 per week per laboratory.

The screening procedure consists of registration; completing a medical history questionnaire; anthropometry; electrocardiography; respirometry; chest X-ray; mammography (for women over 50); retinal photography, tonometry, and a visual acuity test; measurements of blood pressure, pulse rate, and Achilles tendon relaxation time; laboratory examinations of blood and urine; and a tetanus immunization (Collen et al., 1969).

The total cost per patient of this screening is \$21.32. This may be broken down as follows:

Direct costs	\$12.34	
Total unit costs	14.28	
Computer center and		\$21.32 Total cost
data processing	4.50	
Central staff	2.54	

The largest expenditure in the program is for wages and salaries, which account for \$8.60 of the \$12.34 direct costs. The staff for one multiphasic laboratory consists of three receptionists, eight aides, four and a half clerks, six nurse aides, five nurses, five technologists, 25 percent of one M.D. supervisor, 20 percent of one M.D. EKG reader, 20 percent of one M.D. X-ray reader, and 20 percent of one M.D. eye photo reader.

We shall use our model to analyze decision making in the Permanente multiphasic screening program. We define decisions as actions taken by decision makers (not necessarily one person or even one group of persons) that influence both the operations and the goals of the program. This model is

composed of the process, multiphasic screening, and three decision layers—
the search layer, the adaptive layer, and the self-organization layer. The
division into three layers has been done for conceptual purposes only. Each
layer might actually be modeling the same person functioning in different
capacities, or several different groups of people, each with decision responsibilities that do not overlap. We are not specifying the organizational structure of decision making and goal seeking in the multiphasic screening
program. Instead, we are specifying the conceptual structure of this behavior.

The model consists of two main parts—the decision model and the
process (see Figure 2). The process is the multiphasic screening program. In
this model it will be represented by sets of objects and by functions that relate
these sets in some fashion.

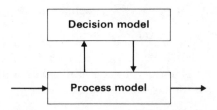

Figure 2. Goal-seeking system model.

Process—multiphasic screening program

The process can be modeled by the following sets and functions:

Time set   t
    indicator that relates time to other objects within the process
    model and the first layer of the decision model

Input Set $X = \{x_t\}$ where $x_t = x(t)$,
    characteristics of the patient population at time t

Output Set $Y = \{y_t\}$ where $y_t = y(t)$,
    information about the patient population generated by
    screening program at time t; the derivation of y_t is shown in the
    outcome function

Decision Set $M = \{m_t\}$ where $m_t = m(t)$,
    decision applied by the decision maker at time t; the derivation
    of m_t will be shown below

Outcome Function

$$P(m_t, x_t) = y_t$$

P takes a single decision and a single input and yields a given output – information about the patient population at time t

$$\downarrow M$$

$$\begin{array}{c} X \\ \longrightarrow \end{array} \boxed{\quad P(m_t, x_t) = y_t \quad} \begin{array}{c} Y \\ \longrightarrow \end{array}$$

PROCESS

Decision model
The decision model is composed of three layers.

Layer one, the search layer, can be modeled by the following sets and functions:

Decision Set $M = \{m_t\}$ as defined above

Output Set $Y = \{y_t\}$ as defined above

Uncertainty Set $U = \{u_\theta\}$ where $u_\theta = u(\theta)$,
  the state of uncertainty concerning the entire system at time θ;
  the derivation of u_θ is shown below

Next Decision Function K

$$K(u_\theta, y_t) = m_{t'}$$

K takes the state of uncertainty of the system during time θ and the last output at time t and yields a decision for the next step at time t'. State of uncertainty is defined as all information about the program that is unknown at any given time

$$\downarrow U$$

$$\begin{array}{c} X \\ \longrightarrow \end{array} \boxed{\quad K(u_\theta, y_t) = m_{t'} \quad}$$

$$\downarrow M$$

Layer two, the adaptive layer, can be modeled as follows:
  Time set θ
        indicator that relates time to objects within the uncertainty set

Uncertainty Set $U = \{u_\theta\}$ as defined above

Decision Set $M = \{m_t\}$ as defined above

Outcome Set $Y = \{y_t\}$ as defined above

Actual Cost Set $\quad C = \{c_\tau\} \quad$ where $c_\tau = c(\tau)$
actual total cost of the program including all nontangible social costs and benefits for period τ

Projected Cost Set $\quad C^* = \{c_\tau^*\} \quad$ where $c_\tau^* = c^*(\tau)$
total projected allowable cost for the program in period τ

Next Uncertainty Function H

$$H(u_\theta, m_t, y_t, c_\tau, c_\tau^*) = u_{\theta'}$$

H takes all of the information generated by the program and the decision maker at time θ, i.e., decisions, outcomes, actual costs, projected costs, and the previous state of uncertainty and yields a new, reduced state of uncertainty for time θ'

$$\downarrow C \qquad \downarrow C^*$$
$$\boxed{H(u_\theta, m_t, y_t, c_\tau, c_\tau^*) = u_{\theta'}} \quad \xrightarrow{\ U\ }$$
$$\downarrow M \qquad \downarrow Y$$

Layer three, the self-organizing layer may be modeled as follows:
Actual Cost Set $\quad C = \{c_\tau\} \quad$ as defined above

Projected Cost Set $\quad C^* = \{c_\tau^*\} \quad$ as defined above

Evaluation Set $\quad V = \{V_{\tau|}\} \quad$ where $V_\tau = V(\tau)$
V_τ is the evaluation function during period τ which determines the actual cost used to evaluate the cost of the program

Evaluation Function V_τ

$$V_\tau(m_t, y_t) = c_\tau$$

V_τ takes a decision and the information generated by the decision into a total cost for the program including all costs *and* all benefits (including nontangibles)

Tolerance Function T

$$T(u_\theta) = c_\tau^*$$

T takes a given state of uncertainty into a projected cost for the time period

Satisfaction Function R

$$R(c_\tau, c_\tau^*) = \begin{cases} \text{satisfaction if } c_\tau \leqslant c_\tau^* \\ \text{dissatisfaction otherwise} \\ \qquad\qquad (R \text{ is self-explanatory}) \end{cases}$$

Next Evaluation Function S

$$S(V_\tau, R(c_\tau, c_\tau^*), u_\theta) = V_{\tau'}$$

S takes the evaluation function during time period τ, the result of the satisfaction function for time τ, and the uncertainty state at time θ into a new value function for the time period τ'

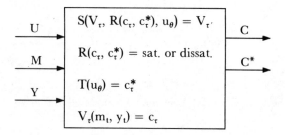

t, θ, and τ are three separate indicators of the passage of time. The interval between t and t' is usually shorter than the interval between θ and θ', which is likewise usually shorter than the length of time between τ and τ'. However, we do not need to specify exactly how much longer the interval from θ to θ' is than the interval from t to t'. It is sufficient to note that these intervals are different. These various time indicators show explicitly that decisions, outputs, uncertainties, and evaluation functions *do* change with time, but not necessarily at the same rate or during the same time period. The total model is diagramed in Figure 3.

The actual operation and development of the Kaiser multiphasic screening program, at the process level, is described by the equation:

$$P(m_t, x_t) = y_t$$

In our simplification of the actual operation of the program we assume that if given a group of operating rules (m_t) at time t and the characteristics of the patient population at the same time, i.e., x_t, the program will produce a given set of information about the patient population, i.e., y_t. This is straightforward.

At the decision level of the model, decision layer one (the search layer) yields the operational rules for the process. We describe this by the equation:

$$K(u_\theta, y_t) = m_{t'}$$

This decision layer takes a given statement of uncertainty— u_θ, which comes from layer two—and the information produced by the screening program, y_t, and produces operating rules for the next time period of the process, $m_{t'}$.

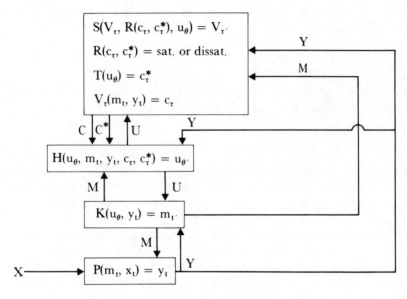

Figure 3.   Total model.

Decision layer two (the adaptive layer) has the task of using all possible sources of information to reduce the uncertainty set and restate it in a new form. The equation

$$H(u_\theta, m_t, y_t, c_\tau, c_\tau^*) = u_{\theta'}$$

describes this task. This layer takes the present state of uncertainty, u_θ; the operating rules; the patient information generated up to time t, m_t and y_t; and both the actual costs and projected costs of running the program, c_τ and c_τ^* sent down by layer three; and yields the new state of uncertainty, $u_{\theta'}$.

The function of decision layer three (the self-organizing layer) is to evaluate the costs and benefits of the program and to develop new, more appropriate evaluation measures as the uncertainty is reduced. This is described by four equations:

$$V_\tau(m_t, y_t) = c_\tau$$

This is the evaluation equation. It takes a given set of operating rules at a given time, m_t, and the information produced by the program at the same time, y_t, and yields a total cost for the operation of the program, c_τ . This cost, c_τ , takes into account all costs and all benefits, both tangible and nontangible, for the entire program.

$$T(u_\theta) = c_\tau^*$$

takes a given state of uncertainty, u_θ, and assigns to it a projected cost for the program, c_τ^*.

$$R(c_\tau, c_\tau^*) = \begin{cases} \text{satisfaction} \\ \text{dissatisfaction} \end{cases}$$

compares the actual cost to the projected cost, c_τ and c_τ^*, and yields either satisfaction or dissatisfaction.

$$S(V_\tau, R(c_\tau, c_\tau^*), u_\theta) = V_{\tau'}$$

takes the present evaluation function at time τ, i.e., V_τ , the result of the satisfaction function during the same period, $R(c_\tau, c_\tau^*)$, and the given state of uncertainty, u_θ from layer two, and yields a new evaluation function, $V_{\tau'}$, for the next time period, τ'.

We should emphasize two points here. First, although we postulated certain equations in our model, this does not necessarily imply that a numerical transformation takes place. That is, while there is no doubt that a strong relationship exists between population characteristics, operating rules, and the information produced by the screening program

$$(P(m_t, x_t) = y_t),$$

we cannot numerically specify the exact nature of the transformation. Second, we use three different time indices, t, θ, and τ, to again emphasize that the three layers operate on three different time scales. These time scales are not fixed intervals, and there is no specification as to the length of time between t and t′, θ and θ′, or τ and τ'. Furthermore, each level operates dynamically in time in relation to the other levels.

We shall attempt to verify our model by describing specific examples from the real world (the actual operation of the multiphasic screening program). As originally conceived and implemented, the goal of multiphasic screening was to detect abnormalities in *asymptomatic* patients—to discover occult, latent, or presymptomatic disease with the hope that early treatment of the discovered conditions would lead to more favorable outcomes (Collen,

1966). This led to the formulation of a set of criteria to evaluate, at least in part, the performance of multiphasic screening. The following four criteria are a simplification of the first evaluation function,

$$V_\tau(m_t, y_t) = c_\tau:$$

(1) the disease or condition should be an important one; (2) there should be an accepted treatment for patients with the disease; (3) there should be an agreed policy on whom to treat as patients; and, (4) the cost of early diagnosis and treatment should be economically balanced in relation to total expenditure for medical care (Wilson, 1968).

Actual operation of the program generated information that then could be used to reduce the uncertainty set, change the evaluation function, or change the operating rules. As an example, mammography to detect early carcinoma of the breast was performed on a group of 12,245 women, most of whom were 40 or more years of age. The unit cost per test was $4.90, not including the costs of central services or data processing. Table 1 (Collen, 1968) shows that not a single case of breast cancer was detected by mammography among the women who were under age 50 and that it cost $24,652 to discover this. All of the positive biopsies among the women in this age group were done for reasons other than a positive mammogram. In addition to the $24,652, unknown amounts were spent in following up the false-positive mammograms (32) and for the nine biopsies that were negative for carcinoma. It cost $5,891 to discover each of the six cases of breast cancer among the women over 50 years of age and in whom the disease was not otherwise detectable at the time of screening. Again, this amount does not include the known cost of following up false-positive results and of negative biopsies.

The results obtained in the study described above indicate that mammography fulfills the first three requirements for early disease detection programs in that a very small number of breast cancers were found before they were detectable by other means. However, this was achieved at a tremendously high cost per positive result and uncalculated but considerable direct and indirect costs of following up the cases of false-positive test results.

This information had a major impact on the Permanente multiphasic screening program. This new information, called y_t, was obtained by using operating rules called m_t. Assuming that the population characteristics, x_t, were stable and unchanging, and using the evaluation function, we find that

$$V_\tau(m_t, y_{t'}) = c_{\tau'}.$$

Thus, according to our four criteria for evaluating the performance of multiphasic screening, mammography does not fulfill criterion number four and yields a much higher cost than we expected. When this new information

	Women less than 50 years old		Women 50 or more years old	
	number	*cost*	*number*	*cost*
Women examined	5,031	$24,652	7,214	$35,348
Positive mammograms	32	$770 per positive result	77	$459 per positive result
Total biopsied	14		33	
Biopsy positive for cancer	5		14	
Cancer with positive mammogram	0	$24,652 to discover no cancers	11	$3,213 per positive case
Cancer and not clinically palpable	0	$24,652 to discover no cancers	6	$5,891 per positive case not detectible by palpation by time of screen
Biopsies with negative results	9	???	19	???

Table 1.   Results of Mammography Testing

is put into the next uncertainty function, H, it yields a new, reduced state of uncertainty, $u_{\theta'}$:

$$H(u_\theta, m_t, y_{t'}, c_{\tau'}, c_\tau^*) = u_{\theta'}$$

The new, reduced uncertainty state is put to two uses. First, it goes to layer one where it is used to devise a new set of operating rules:

$$K(u_{\theta'}, y_{t'}) = m_{t'}$$

In the case of the Kaiser multiphasic screening program, the decision was made to use mammography only for women 48 years of age or older (Collen and Davis, 1969).

The new uncertainty state of $u_{\theta'}$ also goes to layer three. There it yields a new figure for projected costs:

$$T(u_{\theta'}) = c^*_{\tau'}$$

This figure is compared to actual costs through the satisfaction function:

$$R(c_{\tau'}, c^*_{\tau'}) = \text{dissatisfaction}$$

The outcome is dissatisfaction. The next evaluation function therefore yields a new evaluation function:

$$S(V_{\tau}, R(c_{\tau}, c^*_{\tau}), u_{\theta'}) = V_{\tau'}$$

In returning the system to equilibrium, this new evaluation function $V_{\tau'}$ should have the following effects:

$$V_{\tau'}(m_{t'}, y_{t'}) = c^{\wedge}_{\tau'} \text{ and}$$

$$R(c^{\wedge}_{\tau}, c^*_{\tau'}) = \text{satisfaction}$$

That is to say, the self-organizing layer must develop a new evaluation function that makes the program seem satisfactory in terms of its actual costs and benefits compared to the projected costs and benefits.

Let us return to the Permanente program and look at what actually happened. We have already mentioned that the operating rules were changed. Another, more marked change was that the goals (and therefore the evaluation functions) of multiphasic screening were restated. (This change did not arise solely from the mammography data; we simply concentrated on that data to explain the model. Information derived from other tests had similar impact.)

COSTS OF MULTIPHASIC SCREENING FOR CORONARY HEART DISEASE, DIABETES MELLITUS, AND PARATHYROID TUMOR

Coronary heart disease is one of the most troublesome health problems of our time although there is a certain amount of agreement on whom to treat and how to treat them. The cost in dollars per positive test for four coronary heart disease tests is (Collen et al., 1970):

	Unit Cost	% Positive	Cost per Positive
Electrocardiography	1.02	17.3	5.90
Blood pressure	.42	4.1	10.40
Serum cholesterol	.29	2.4	12.15
Serum uric acid	.29	4.5	6.40

These costs do not by themselves appear excessive, especially when one considers that in 1963 the total cost of treating diseases of the circulatory system was estimated to be between 19 and 22 billion dollars, which was more than 22 percent of all costs for all illnesses treated (Rice, 1966).

The clinical manifestations of coronary heart disease are angina pectoris, coronary insufficiency syndrome, myocardial infarction, sudden unexpected death, and congestive heart failure. While none of these conditions can be cured, all except sudden death can be treated. Screening for ischemic heart disease is best accomplished by questionnaire, rather than by chemical or physical procedure (Fodor, 1968). But can multiphasic screening be useful in detecting persons who are highly susceptible? The answer is a definite yes. Abnormal results on any of the four tests mentioned above correlate well with an increased likelihood that the patient will develop overt, symptomatic disease. Similarly, positive findings in two or more of these tests indicate the presence of greater hazard than in any one test alone. But two factors that are tremendously important in determining future susceptibility to coronary heart disease are cigarette smoking and obesity. Whether a person who is identified as being in the susceptible group can be treated is not clear. Usually the first step in treatment is to require that he quit smoking and lose weight. This "medicine" is easy to prescribe but very difficult to take. Similarly, the most satisfactory kind of long-term control of hypercholesterolemia (and associated hyperlipidemias) and of moderate hypertension is dietary control. Medications are of dubious value in the former, and are costly and perhaps dangerous in the latter. Currently, the physician does not treat an asymptomatic patient for abnormalities in the electrocardiogram or for hyperuricemia that does not produce gout.

In summary, while this screening fulfills the four requirements for early disease detection programs mentioned previously, treatment for the presymptomatic or susceptible state is either nonexistent or nonacceptable to the population at risk. Many people have quit smoking cigarettes, but others continue, and obesity is the single most common abnormal physical condition in adults in this country (Sanders, 1964).

At first glance, diabetes mellitus, an important disease that affects a significant number of people, might seem the ideal disease to use for justifying multiphasic screening, using the original evaluation function. The unit cost for a one-hour glucose determination (including challenge dose) is $0.75 and the overall yield of 5.7 percent positive results brings the unit cost per positive test to $13.15. A two-hour test is then done on the one-hour positives, and this yields a final 3.5 percent positive result, for a unit cost of $19.40 per positive diagnosis (Collen et al., 1970). There is a treatment for the disease as well as general agreement as to which patients should be treated. What is lacking is substantial evidence that medical treatment for adult-onset, nonketotic diabetes does any good. Usually the best treatment for this type of diabetes is dietary. The pharmacologic forms of treatment presently available can normalize blood sugar levels, but they have not been shown to delay or prevent the pathological processes of this disease including damage to the eyes, kidneys, blood vessels, and nerves. Presently, there is no satisfactory evidence that patients treated vigorously do better than those not treated at all. On the contrary, preliminary results of a long-term study of the oral "antidiabetic" drug tolbutamide show increased mortality in the treatment group compared to the control group.

While the unit cost of a serum calcium determination is only $0.29 and the incidence of positive tests is 1.3 percent, the incidence of parathyroid tumors is only 0.02 percent, yielding a cost of $1450 per diagnosis of parathyroid tumor, plus the uncalculated cost of following up false-positive tests (Collen et al., 1970; Collen, 1969).

Similar arguments can be offered in regard to tonometry as a screening test for glaucoma, chest X-rays and respirometry for pulmonary disease, hemoglobin determinations for iron deficiency anemias, serum transaminase determinations for liver disease, and urine cultures for asymptomatic urinary tract infections (Garfield, 1970a).

To summarize, the evaluation of multiphasic screening solely on the basis of the detection of asymptomatic disease has been discouraging. It costs the Permanente Group over one million dollars a year to run the program, and this does not include the cost of follow-up of the many false-positive results. Furthermore, treatment for many of the conditions discovered is either ineffective or unacceptable to the population. When faced with this information (a reduction of the uncertainty set), the decision makers at the Permanente Group had several choices, one of which was to terminate the program. But such a decision would have involved dislocation and replacement costs which also had to be taken into account. Actually, the decision made was the one predicted by our model. The change in the uncertainty set led to a marked change in the evaluation function. The goals of multiphasic screening, rather than concentrating solely on the detection of asymptomic disease, were expanded to include the several modifications. These goals were not formulated at once, but in an evolutionary manner as new information was accumulated from the experience of running the program (Collen, 1969; Garfield, 1970a, 1970b).

    1.   The refined goals included using the program as a gate control for

patients entering the delivery system, including separating the well from the sick; establishing entry priorities for the sick; and providing a preliminary workup for the sick.

2. New goals that were added included using the program to assist in the diagnostic process to save physicians' time by reducing the number of patient visits; to help fulfill the demand for preventive health services by optimizing the use of allied health care personnel; as a tool for medical research, by evaluating new tests, discovering new epidemiologic relationships, and establishing more correct norms; and finally, to establish a health profile for continuing care, i.e., a computerized data base.

We shall now examine each of these refined and new goals in turn. The use of multiphasic screening as a gate control for patients entering the delivery system seems a reasonable procedure. The present system (outside of the Permanente Group) uses physicians as gate controllers. When a patient is sick or thinks he is sick he goes to see a physician. The physician decides on the next step: (1) he obtains a workup for diagnostic purposes; (2) he initiates an appropriate treatment regimen if a diagnosis is established, or reassures the patient if no evidence of disease is found (see Figure 4).

It is obvious that multiphasic screening separates the well from the sick (at high cost and subject to the limitations discussed earlier), helps establish

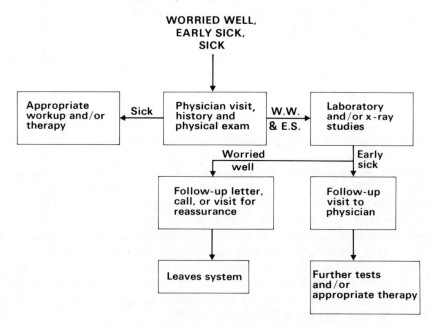

Figure 4.   Present system with physicians as gate controllers.

entry priorities for the sick, and provides a preliminary workup for the sick (Figure 5). The relevant question, however, is whether it performs in these functions better than the present system (a use of the evaluation functions); the answer is uncertain. Given the alternative of three hours of multiphasic screening or ten to fifteen minutes with a physician, most patients will choose the latter. Multiphasic screening comes close to being the ultimate "shotgun" approach, whereas a physician can be much more selective and specific in the tests he orders. While the cost of these individual tests may be higher than when performed in a multiphasic program, unnecessary tests are ordered less often. We presently have no data concerning laboratory and X-ray costs for an *average* first office visit to a physician, but it is very unlikely that it approaches $21.32, the cost of multiphasic screening (Collen et al., 1969).

No data are currently available regarding the performance measure of multiphasic screening in assisting the physician to make his diagnosis. However, it is likely that diagnoses can be made and/or confirmed on the basis of fewer and more selective laboratory examinations at less than the cost of multiphasic screening.

A related goal of multiphasic screening, that of saving time for the physician, is a moot subject. It definitely does not save time for the patient. Time saved for the physician is a result of the reduction in the number of patient visits. This difference occurs primarily when the group of patients seen are in the early-sick category. In both the present health care system and

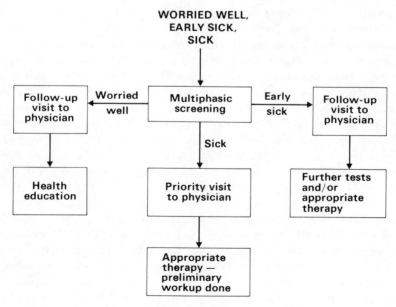

Figure 5.   Multiphasic screening as gate control.

the multiphasic screening system those who are sick go directly into therapy following the first physician visit. In the present system, the worried-well most often have only physician visit with follow-up and reassurance being handled by letter or phone, but the early-sick patient must have a second visit to initiate further workup and/or therapy. In multiphasic screening workup and initiation of therapy can be accomplished during the first physician visit. However, when the treatment is long and protracted, the time saved by eliminating one visit loses significance. Finally, it is necessary to take into account the time required for the physician to follow up false-positive test results generated by multiphasic screening.

Again, it does not seem that the evaluation functions for the three refined goals for screening yield particularly favorable results, but the data here are not as good, i.e., the uncertainty set is still too large to tell us much. The Permanente Group is presently attempting to generate more concrete performance data relating to the advantages and disadvantages of screening as a gate control mechanism.

The most recently proposed goal for screening is the effective organization of our resources to meet the demand for preventive health services. As health care has become recognized as a right rather than a privilege, demands for both preventive services (checkups) and medical treatment for illness have increased. Since we lack sufficient physicians to satisfy both of these demands, we must use nonphysicians to provide health services for those who are not sick. Multiphasic screening helps fulfill the demand for preventive services, partly by using nonphysician personnel almost exclusively.

Granted that the demand for preventive services exists, the critical question involves selecting the preventive services to which we should allocate our resources and nonphysician manpower. As stated earlier, we have little evidence that multiphasic screening is an efficient way of detecting asymptomatic disease. But we have even less evidence to show that periodic "checkups" accomplish this purpose. However, preventive services must be provided and must be ones that can demonstrate a beneficial effect upon the health of the population. Screening does use nonphysician manpower effectively, but might not other preventive services utilize this manpower even more effectively? Perhaps the single most beneficial preventive service that could be offered is *effective* health education — health education that is substantive rather than imaginative. If such a program could educate people to stop smoking, to eliminate obesity, and to exercise regularly, a demonstrable change in the health status of the population would result. If we could educate people to make better use of the existing health care delivery system, we could show positive results. A far greater need than the separation of well and sick persons by some gate mechanism, it would seem, is that of getting some people with *symptomatic* illness into the medical care delivery system at all. Perhaps better health education could remove some of the barriers to care that are caused by ignorance and fear. Another benefit of health education could be improved patient performance in following physicians' directions, since it is well known that a great deal of medical care is negated because of failure to follow a prescribed therapeutic regimen.

Still another goal for screening is that it be used as a tool for medical research. Comprehensive screening of large numbers of people can give us epidemiologic information about populations. It can help us to make test results more meaningful by establishing more accurate and correct norms and can lead to the discovery of new relationships between the variables involved. It can also give us information about how those variables change over time and provide us with a standard against which we can compare new tests and procedures. Certainly multiphasic screening generates information that can be useful in medical research. The problem is how to calculate the value of this benefit. The cost analysis of any research is a complex procedure that has not yet been satisfactorily carried out, and that for medical research is extremely complex.

Finally, it has also been proposed that screening establishes a health profile for continuing care, a computerized data base. In our opinion, this is the most important goal the program has. Medical record-keeping, which for the most part is still at a very primitive level, is the source of one of the major problems in medicine—information retrieval. The cost of unnecessary duplication, inaccurate information, lost information, and poor organization of information is certainly very high. The saving grace of multiphasic screening is that it provides a data base for an individual that is tremendously useful to physicians because it is standardized and can be processed by computer. It can help provide a complete and accurate medical history including both subjective and objective data. Such a data base, generated by screening and other contacts with the system, could be periodically updated (research could tell us how often) and could be nationally standardized so as to be available no matter where an individual happened to be. This is certainly within our capabilities, both technologically and economically. The stumbling blocks are essentially political and will probably be overcome as the government becomes more involved in the financing and delivery of health care. The attractiveness of this prospect is that the costs would go down and the benefits would go up as such a program became more widespread.

SUMMARY

In this paper we have followed the development of multiphasic screening in the context of a model of a decision-making, goal-seeking system. We have pointed out that as the program progressed and generated information to reduce the uncertainty set, certain changes became necessary in operating procedures, goals, and evaluation functions. The gradual development of new goals in any system follows a distinct pattern. When a new goal is proposed, the evaluation functions are correspondingly altered. As information is acquired to reduce the uncertainty set further, the need for different goals may become apparent. This was the case for the goal in our program of multiphasic screening for early detection of disease. It also may turn out to be the case for the related goals of acting as a gate control, assisting in the diagnostic process, saving physician time, and fulfilling the demand for preventive health services. However, the information in regard to these latter goals is not as available or as precise as it was for the original goal, and we are left with too much uncertainty to say for sure just what will happen. But we do know what kind of further information we want. The newest goals for screening—the use of the screening program as a tool in medical research and in establishing a computerized health profile—have the least evaluative information presently available. A study is now underway to reduce the uncertainties relating to these two goals. Perhaps as this information becomes available, we shall again have to reformulate the goals for the program.

We have introduced the concept of a multilayer, goal-seeking system model and used it to explain the evolution of goals for multiphasic screening; a simple input-output model would have been inadequate. Although a model such as we have presented was not essential to the above analysis, it did provide a conceptual framework within which we could organize the diverse information confronting us on a more meaningful fashion—which is the value of any theory or conceptual framework.

REFERENCES

Collen, M. F. Periodic health examinations using an automated multitest laboratory. *Journal of the American Medical Association*, 1966, *195*, 830-833.
——. Automated multiphasic screening. In C.L.E.H. Sharp and H. Keen (eds.), *Presymptomatic Detection and Early Diagnosis*. London: Pitman, 1968.
——. Value of multiphasic health checkups. *New England Journal of Medicine*, 1969, *280*, 1071-1073.
Collen, M. F., and Davis, L. F. The multitest laboratory in health care. *Occupational Health Nursing*, 1969, *17*(7), 13-18.
Collen, M. F., Kidd, P. H., Feldman, R., and Cutler, J. L. Cost analysis of a multiphasic screening program. *New England Journal of Medicine*, 1969, *280*, 1043-1045.

Collen, M. F., Feldman, R., Siegelaub, A. B., and Crawford, D. Dollar cost per positive test for automated multiphasic screening. *New England Journal of Medicine*, 1970, *283*, 459-463.

Fodor, J. Screening for ischaemic heart disease. In C.L.E.H. Sharp and H. Keen (eds.), *Presymptomatic Detection and Early Diagnosis.* London: Pitman, 1968.

Garfield, S. R. The delivery of medical care. *Scientific American,* 1970, *222*(4), 15-23. (a)

――――. Multiphasic health testing and medical care as a right. *New England Journal of Medicine*, 1970, *283*, 1087-1089. (b)

Mesarovic, M. D., Macko, D., and Takahara, Y. *Theory of Hierarchical, Multilevel Systems*. New York: Academic Press, 1970.

Rice, D. P. Estimating the cost of illness. *Health Economics Series No. 6*, Public Health Service Publication, Washington, D.C., 1966.

Sanders, B. S. Measuring community health levels. *American Journal of Public Health*, 1964, *54*, 1063-1070.

Wilson, J. M. G. The worth of detecting occult disease. In C.L.E.H. Sharp and H. Keen (eds.), *Presymptomatic Detection and Early Diagnosis*. London: Pitman, 1968.

PART FOUR

Designing Research Using Systems Theory

Previously described health care subsystems have demonstrated varying degrees of relationship to patient care. Currently, one model in which the patient is an integral element of the system is being developed under Howland and co-workers at Ohio State University. These researchers have been working with a health system model that they hope will show how strategic resource allocation decisions can affect patient care. Considerable impetus has been given to the work of this group by recent federal legislation (PL 91-515) which specifically states that systems analysis methods will be used to develop plans for health care systems. Howland reiterates in his paper the questions that many concerned observers have raised on the ineffective and expensive health care system now available. He also proposes that the most crucial problem in using the concept of systems planning relates to the validity of the models that are used. He and his colleagues set themselves the task of constructing a model by using an inductive synthesis method to ensure that they are "telling it like it is." Their work has progressed slowly, in part because of this methodology.

One of the major problems Howland confronts is that systems analysis, as he sees it, generally focuses on resource allocation, using a return-on-investment criterion. To provide a way of relating resource allocation to utilization, which is the primary concern of the receiver of the health service, the Ohio State investigators partition the system into three interacting levels: strategic, operational, and tactical. The cybernetic, adaptive systems model has been developed at the tactical level. Convinced that the nurse-patient-physician triad is the primary functioning unit of the health system, the researchers began their investigation with that triad.

Howland's presentation of the process of delineating the criteria for health system performance might raise some controversy. There are some who believe that the complexity of health care does not lend itself to placing full responsibility upon one person, even though that person is a well-qualified physician. There is increasing emphasis on the need to devise a collaborative system with shared responsiblity and to integrate the skills of a variety of health care professionals on behalf of the patient. However, this group of researchers were satisfied to accept the hypothesis that patient care can be measured in terms of the system's ability to maintain selected patient variables within physician-prescribed limits.

Using Ross Ashby's three basic steps as outlined in An Introduction to Cybernetics *(Ashby, 1956), the researchers developed a descriptive model of the tactical level of the system. The collection of the "generous length of record" is the first of these three steps. The researchers collected time-series data, using ten patients who had had uncomplicated abdominal hysterectomies. Observation of patients took place for 24 hours a day from the day of admission to the day of discharge. Pierce's paper presents in detail the methods used in collecting the data and the problems encountered because of the nature and amount of data. Analysis of the data is currently taking place.*

LaRue discusses the computer applications and extensions. The possible use of graphics in the display of health information is an intriguing development. The second and third steps, which involve examining the record, can be operationalized most efficiently with the computer applications.

In these three papers, the research group proposes that a clear picture of utilization of the systems resources as they are brought into play to maintain the prescribed boundaries of the variables at this tactical level can be useful at the strategic and operational levels of the system. The group is continuing its work with analysis of the data toward that goal.

Daubenmire examines the nature of the nurse-patient-physician triad communicative interaction system, on the basis of her work with the Adaptive Systems Research Group. She outlines a methodology to examine the nurse-patient-physician communicative interaction process by describing patient state, nurse state, and physician state at a given point in time, and she predicts that insight into essential problems in nursing will result. The interaction of the participants moving toward goal achievement can be observed through the changing states of the individuals as they are affected by the messages transmitted in the communicative process. The results of Daubenmire's study have important implications for nursing intervention. She conducted a pilot study to test the feasibility of collecting data at various times during a patient's hospitalization and here confronts the problem of recording nonverbal data. One of the interesting resources she used in her attempts to gain understanding of nonverbal communication has been the study of dance.

Buckeridge is using the Adaptive Systems Model in her current study of nursing care of the patient in neurological stress. Data collected in a pilot study is presently being analyzed so that relationships between components of the patient care system can be identified and testable hypotheses can be proposed. Her paper details the process of data analysis. Subsequent phases of her research will include an exploratory study to identify the essential information the professional nurse needs in this area of patient care. Buckeridge hopes to identify nursing content for all levels of nursing personnel engaged in caring for the patient in neurological stress.

Sills outlines some of the difficulties encountered in designing systems theory research projects that have the potential of affecting practice in a meaningful way. She implies that consideration of some of the factors involved in the creation of *the gap between systems theory and research could lead to improved communication between theorists and practitioners and, subsequently, to more refined theory. Her belief that the alternative is hopeless desperation for the future of civilization might be overdrawn, but a number of prestigious thinkers hold this position.*

CHAPTER 9

An Adaptive
Health System Model

Daniel Howland

Growing concern with the rising costs and demands of health services has produced intensified political action to ensure that the health needs of society are met. The problems that have arisen with the operation of Medicare and Medicaid, and current proposals for replacing these plans with alternatives, require a better understanding of the functioning of the health system. Public Law 91-515 (1970), for example, specifies that the "systems analysis methods will be used" to develop plans for health care, to ensure that these systems are "designed adequately to meet the needs of the American people." The requirement for planning is clear. Less clear is whether or not the conventional methods of analyzing systems will do the job.

The Department of Defense's experience with systems analysis and planning-programming-budgeting systems raises questions about the methods effectiveness in providing information needed for the solution of health system problems. How, for example, are system goals to be specified? How are costs and benefits to be measured? Health is an individual matter, and measures of effectiveness must reflect the impact of the allocation of health system resources on the individual, not on groups of people. What is true for the group may be false for the individual. No family, for example, can have 2.5 children, although this can be the average family size for a group.

Serious questions have been raised about the rationale underlying systems analysis methods and the way they have been used. Some of these criticisms seem irrelevant, such as concern about development of "whiz kids" and the use of computers. Other criticisms, such as the omission of social and behavioral variables from the models used for analysis, must be remedied if the methods are to provide the information required to solve broad societal problems.

The criticisms of the planning methods have been discussed at length by proponents and adversaries before the Jackson and Proxmire committees of the U.S. House and Senate (U.S. Congress Subcommittee on Economy in Government, 1969; U.S. Senate Committee on Government Operations, 1970).

An examination of the criticisms suggests that it is *not* the concept of systems planning which causes concern. There is no argument with the desirability of specifying goals and alternative ways of achieving them, costing the alternatives, and selecting the one that most nearly achieves the goals. Problems arise, however, when real-world predictions are inaccurate and the systems are unable to meet cost and effectiveness criteria. If model-based planning methods are to be used, they must provide valid and relevant information. To meet these criteria, the models used for analysis must be descriptive of the systems for which decisions are being made.

Systems analysis, as developed in the Department of Defense, is a top level, strategic activity aimed at the solution of resource allocation problems (Enthoven, 1964). It looks to the future and provides guidance for decision makers faced with problems of allocating resources to unbuilt systems. Resources are allocated to both cost and "effectiveness" criteria.

From the standpoint of the patient in the hospital, however, the problem is not resource allocation but resource utilization. As far as the individual patient is concerned, resources have already been allocated and what counts is how they are used. It is imperative, therefore, that any model used for analyzing health systems relates resource allocation at top community levels to resource utilization at the bedside. Strategic decisions on allocation should be based not only on costs, but also on effectiveness, measured in terms of the quality of care provided for the individual patient. Strategy depends on tactics, and unless tactical capabilities and limitations are understood by the strategic planners, strategic requirements may not be met. It is impossible to provide care for a population without providing it for the individual members of the population. Our research objective, therefore, is to develop and validate a hierarchical set of models that can be used to trace the consequences of resource allocation from the strategic allocator to the tactical utilizer, the individual patient.

To develop such a set of models we have partitioned the system into three interacting levels (Howland, 1964, 1970; Figure 1).

Starting at the tactical level, we are basing our model hierarchy on a description of how resources are utilized at the bedside. The resulting set of

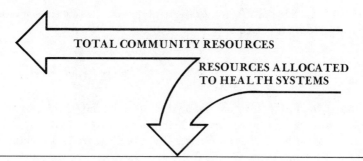

STRATEGIC LEVEL
Systems Analysis

Long-range planning of future health facilities:

What future demands are expected?
What facilities will be required to meet them?
How will these facilities be financed?

OPERATIONAL LEVEL
Operations Research

Management of groups of patients in an existing health facility:

What demands are expected for groups of patients?
What data about individual patients are required to
develop this information?
What resources will be required to meet the demands?
How will these resources be acquired?

TACTICAL LEVEL
Adaptive Systems Research

Management of the individual patient:

How does the system adapt to patient demands?
How does the use of a specific resource affect a
given patient?

Figure 1.   The functional levels of a health system.
(From D. Howland, "Toward a Community Health System Model," in
Systems and Medical Care, ed. A. Sheldon, S. Baker, and C. P.
McLaughlin. Copyright M.I.T. Press, 1970)

models will make it possible for strategic planners to assess the consequences of their decisions for patients.

THE CONCEPTUAL VIEW

The underlying rationale of systems analysis is economic (Hitch and McKean, 1960). When choices must be made between scarce resources, trade-off relationships provide a basis for choice. Given realistic trade-off relationships, based on valid measures of effectiveness and realistic costs, one can make decisions to optimize specified utility functions. Problems arise, however, in specifying the utility functions. A major problem is the assignment of monetary values to social, psychological, and physiological factors such as human life. As Hitch and McKean (1960) point out: "there is no generally acceptable method of valuing human lives" (p. 185).

Partly in an effort to avoid this problem, and partly because a wide range of systems, including health systems, behave in an adaptive manner, we have selected cybernetic models as representative of this type of behavior. We believe that the adaptive notions of biology and physiology provide a more useful rationale for system modeling that do the bases of microeconomics. As a result, we look to Bernard (1865) and Cannon (1932) rather than Adam Smith and Jeremy Bentham for a conceptual framework for our work.

The methodological considerations underlying our research include:

1. *Starting with tactics.* To provide information for resource allocations at community levels, one must understand tactical behavior. It follows, therefore, that modeling must begin at the tactical level. The alternative is to assume tactical behavior, which is easier than measuring it. The assumptions, however, may be wrong. Plans at the community level, for example, are usually based on demand and resource availability data, but the form of the demand distributions depends on tactics. We have begun model development at the tactical level to determine how resource utilization affects these distributions.

2. *The conceptual framework.* At the tactical level we have conceptualized system behavior as adaptive and have selected cybernetic models to represent it. We are concerned with problems of stability in time, rather than problems of optimizing in relation to utility criteria. Both concepts are useful but for different purposes and at different levels in the system. Adapting at the tactical level is a matter of resource utilization; at the strategic level it is a matter of resource allocation, and concepts of return on the investment are relevant.

3. *Dealing with complexity.* Because of the complexity of health systems, and our ignorance about the way they behave, we are attempting a

descriptive rather than prescriptive approach; description must precede prescription. The time-varying behavior of large systems, abstracted as a "list of variables" (Ashby, 1956, p. 39), can be analyzed. The concomitant variation of the variables in time can then be described as changes in the values of system state vectors. Computer programs are available to analyze these changes.

4. *Formal and empirical methods.* As a consequence of our insistence on descriptive models to provide a basis for prescription, our research methods combine empirical methods based on observation and measurement with formal methods based on axioms and postulates. The empirical phase must precede the formal to clearly establish what is to be formalized.

5. *The data base.* Since we have started our modeling at the tactical level to develop the trade-off relationships required for the application of economic analysis at the strategic level, our data comprise direct observations of system behavior in time. We need the resulting time-series data to understand how the system uses its resources to cope with disturbances. From it we can develop time-sampling plans to conduct statistical analyses.

6. *The change agent.* There is a great deal of talk about the need for change in health-care systems, but there is little agreement about what to change and how change can be implemented. We assume that information itself is the most powerful change agent in the evolution of a system. If we can specify the criteria for system performance and measure the ability of a system to meet these criteria, we can expect public pressures to force the system to change to satisfy these criteria. This can be observed happening today as health systems change to provide care for segments of the population which heretofore were not much involved. This assumption presents an interesting paradox. If information is truly a change agent, and if a system will adjust to conform to society norms as its behavior is better understood, we move toward a situation of better and better information. One can then speculate on the limits. Can the free-enterprise system, as we know it, operate in a noise-free information environment, or do we face a situation in which the public sector operates in a high information environment and the private sector in a low information environment? In any case, we can probably expect a gradual shift of health services from the private to the public sector with an accompanying increase in information.

CYBERNETIC MODELING

The methods we are using are based on general concepts described by Ross Ashby in his remarkable books *An Introduction to Cybernetics* (1956) and *Design for a Brain* (1960). In the former he outlines three basic steps in the modeling of complex systems: (1) obtain a generous length of record—every system, fundamentally, is investigated by the collection of a long protocol, drawn out in time, showing the sequence of input and output states, (2) look for patterns of behavior in the record, and (3) express the regularities as transformations.

Before describing the methods we have developed to collect data, search for patterns, and construct transformations, we should define key terms.

Patients are defined as members of a society who are unable to lead what Bernard (1865) called "the free and independent life" because of lack of stability in the internal environment; that is to say, they are people whose natural homeostatic mechanisms have broken down (Howland and Mc-Dowell, 1964). Questions can immediately be raised about patients who will never be able to lead the "free and independent life." These include para-plegics, polio victims, and terminal patients. How do we establish criteria for the effectiveness of system responses to their needs? If we then define health systems, and most especially hospitals, as the organizations which undertake to restore or maintain the homeostasis of the internal environment when a patient is unable to cope by himself, the criterion for hospitals or health system performance is the degree to which variability in patient behavior can be maintained within specified limits. We are not, at this stage in our research, concerned with the mechanism for specifying the limits. Rather, given limits arrived at by whatever means, we wish to find out what resources and operational rules are used to hold patient variables within them.

We can examine our data to find out what limits actually were main-tained. Thus, in the case of the terminal patient, the responsible physician may judge some variables to be uncontrollable, and he may elect to focus his attention on a selected few, such as pain and anxiety. The system would not be judged with respect to the impossible criterion of keeping the patient "free and independent." Rather, it would be judged with respect to its ability to respond appropriately to the patient's demands as perceived by the respon-sible physician. Patient care can then be measured in terms of the health system's ability to maintain selected patient variables within prescribed limits.

We are representing tactical level behavior with a cybernetic model, (Figure 2). In this model we can relate resource allocations to patient demands. We can assign resources and determine relationships between resource utilization and levels of homeostasis. System effectiveness is then measured in terms of the system's ability to maintain the patient in desired states, an easier thing to measure than utility.

Since we have elected to develop a cybernetic model to describe the behavior of an adaptive system, we must collect and analyze time-series data to show the concomitant variation of patient and resource variables in time. We collected data showing the interactions of patient and members of the health team in time and plotted them by computer (LaRue, 1969). These plots (Figure 3) provide detailed information on the allocation of resources to patients and patient responses in time.

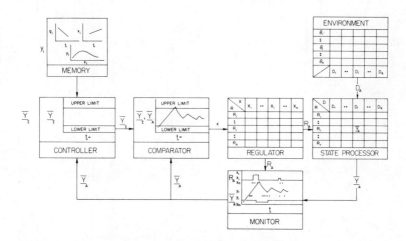

Figure 2. Tactical model.
(From D. Howland, "Toward a Community Health System Model," in *Systems and Medical Care,* ed. A. Sheldon, S. Baker, and C. P. McLaughlin. Copyright M.I.T. Press, 1970)

Having developed a real-time picture of the interactions between the patient and the system, we can pinpoint events of interest in the record. The next question is: How are events dealt with? What events preceded a resource utilization, and how did the patient respond? A computer program has been developed to search the data forward and backward from any specified point in time and tabulate the values of the vector states preceeding and following an event, such as a dangerous drop in blood pressure. I am using the word "vector" in its mathematical rather than its epidemiological sense. It does not mean the carrier of a disease, such as a mosquito, but rather represents the values assumed by a selective set of variables, such as blood pressure, pulse, temperature, and respiration, in any specified time period. The search program (Reid, 1970) makes it possible to find out what resources are used to cope with disturbances and how effective they are in driving patient variables back within limits. For example, when a patient complains of pain, we would like to know whether there are any other signals that would make it possible for the nurse to decide if the patient needs medication to modify a physiological state or needs a conversation to mediate a psychological state.

The third aspect of the analysis is the development of a "Regulator-

PATIENT BEHAVIOR (Y)

SYSTOLIC BP — 150 100 50
DIASTOLIC BP
RESPIRN RATE — 30 20 10
PULSE RATE — 150 100 50
TEMPERATURE — 104 100 98
RSPNS VERBAL
RSPNS PAIN
PUPIL REFLCT
EYES
SQUEEZE
PICKS
MOVES
LIES
FLEX
COUGH

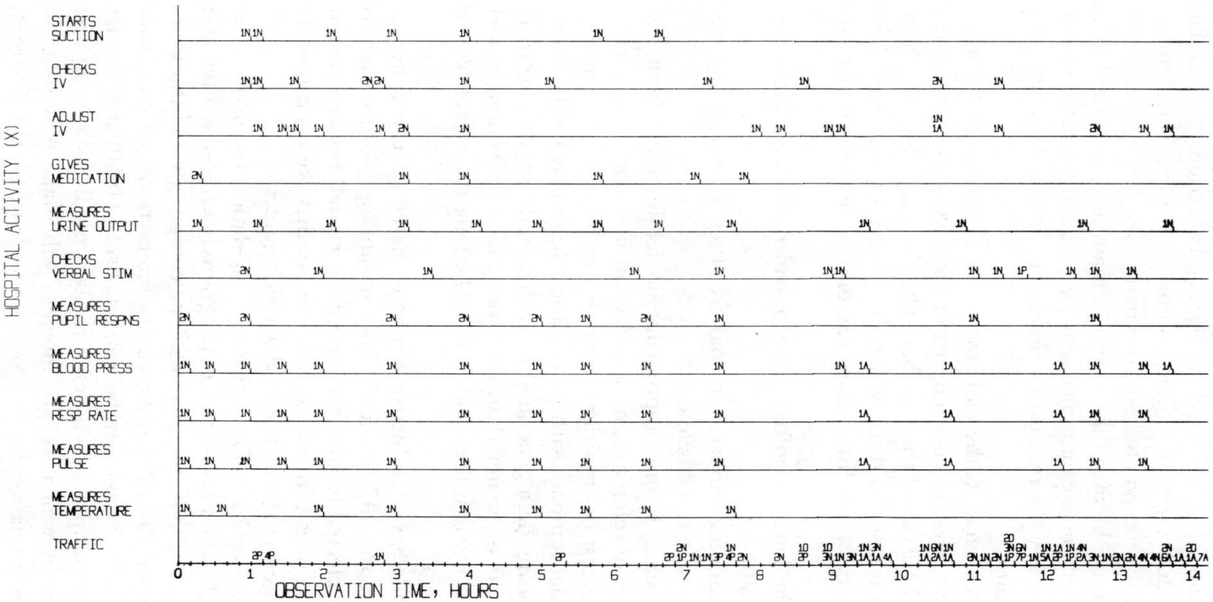
HOSPITAL ACTIVITY (X)
STARTS SUCTION
CHECKS IV
ADJUST IV
GIVES MEDICATION
MEASURES URINE OUTPUT
CHECKS VERBAL STIM
MEASURES PUPIL RESPNS
MEASURES BLOOD PRESS
MEASURES RESP RATE
MEASURES PULSE
MEASURES TEMPERATURE
TRAFFIC
0 1 2 3 4 5 6 7 8 9 10 11 12 13 14
OBSERVATION TIME, HOURS

Legend	Hospital Personnel Performing Activities:	
1, 2 . . . 9 frequency of activity	A = Nurse Aid	
x = eyes closed	L = Licensed Practical Nurse	O = Orderly
O = eyes open	N = Registered Nurse	P = Physician

Figure 3.
Continuous time-series plot showing concomitant variation of patient and resource variables of a 14-hour period.

Disturbance" table (Ashby, 1956, p. 202), which can be used to provide guidance with respect to the expected outcome of system interventions. This table shows the states of the "state processor," the patient, following a disturbance and the utilization of a resource, a regulatory action. It can be developed by summing data for each patient and for all patients with similar disturbances. It has, Ashby has pointed out, two main uses. If the regulator is able to see the disturbance, he may take action to maintain the patient in a desired state. If the disturbances cannot be seen, the regulator can infer them by looking at the patient and the results of a regulatory action. Many clinical problems are of this nature.

The table illustrates what Ashby calls the "Law of Requisite Variety" (1956, p. 206). He defines variety as the number of states the system can be in, or the logarithm of the base two of this number. The law states that variety in the state of the "state processor" is dealt with only by variety in the regulator. That is to say, complexity in the disturbance can only be dealt with by complexity in the regulator.

Figure 4 details the kinds of information provided by analyzing time-series data in our program.

This method of building models for analysis raises a number of problems, some of which are discussed below.

1. *The cost of data.* If one approaches model building from the formal, deductive frame of reference, data collection may not be necessary, and a large part of the cost of empirical research can be avoided. The risk involved, however, is that a model with no empirical basis may not be valid, and decisions reached using data generated by such a model may be in error. If the cost of these errors is greater than the cost of data collection, model building, and analysis, clearly it is economical in the long run to collect the data and build the models.

2. *Lack of methodology.* Inductive synthesis, based on experimental methods, may be a more difficult and time-consuming process than is deductive analysis. We elected to construct empirical models because we did not feel that the formal ones with which we were familiar were sufficiently descriptive of health systems to provide useful decision guides.

3. *Reliance on computers.* The lightning calculation features of the computer can be brought into play only after programs have been written and debugged and data are available for analysis. The computer's usefulness lies in the fact that numbers of variables and complex interactions, which cannot be examined by any other method, can be studied. Ashby (1956) and Wiener (1948), did not have the computer facilities of today. I predict that computer graphics, the direction in which we are moving, will make it possible to analyze data and discover relationships on an unprecedented scale. There is, of course, the danger of undirected or misdirected fishing expeditions; but, if the data are examined with specific questions in mind, computers will provide a capability never before available for examining relationships.

PATIENT CARE QUESTIONS

1. How are resources allocated in time?

   How do patients respond to resource allocations in time?

2. How does the individual patient respond to resource allocations?

3. What events precede and follow the allocation of a resource?

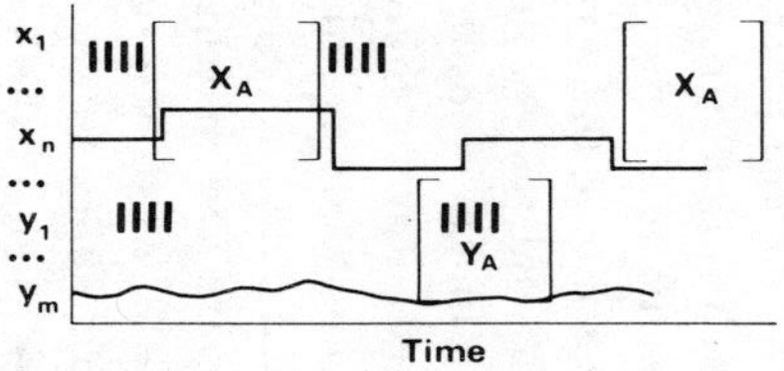

4. How can patient-system interactions be summarized across patients in a probabilistic model?

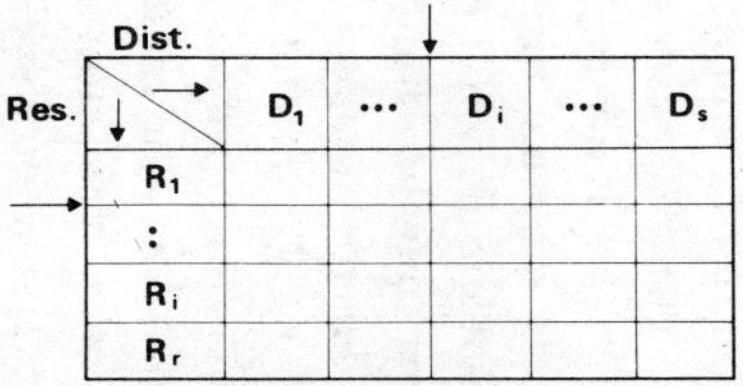

INFO. TO ANSWER QUESTION

Frequency of resource allocation.

Frequency of patient response

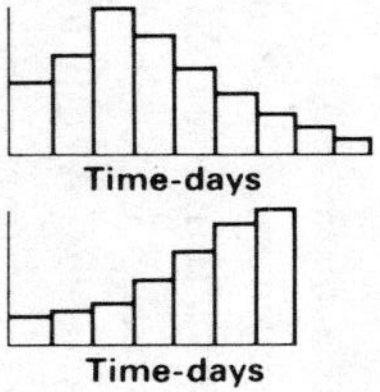

COMPUTER PROGRAM

Program to cross-tabulate the frequencies of resources allocated or patient behaviors for a specified time interval and print these results as histograms.

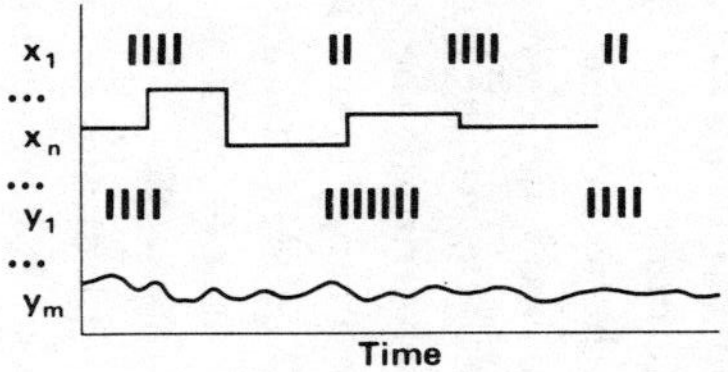

Continuous time series plotting program which displays values for both patient state (y) and resource (x) variables.

Search program to locate specified patient state (y) and resource (x) vectors following and preceding specified events.

A program to sum the x and y vectors from the search program across patients. Cell values represent the state of the patient following a disturbance (D) and regulatory action (R).

Figure 4.   Information and Program Summary

119

4. *Elegance.* In some ways it seems ridiculous to talk about the problem of elegance, but it is a valid problem. It may be conceptualized as the problem of technique versus problem orientation.

An aura of respectability has long been attached to mathematical methods, and the empiricist who insists upon observation, measurement, and experiments, i.e., inductive synthesis, has been regarded as something of a scientific peasant. Mathematical expressions have provided guidance for the solution of problems in many areas. However, we are now talking about an area in which the simplifying assumptions that must be made to ensure mathematical tractability may eliminate the real world problems. By assuming linearity and independence, for example, one may assume away major operational problems. If, however, one assumes a problem orientation, the question of elegance is irrelevant and one can safely use any appropriate methodology.

5. *Enshrining the conventional wisdom.* A criticism often leveled at the empirical approach is that it simply enshrines the conventional wisdom. Conventional wisdom, however, may be more related to *how* things are done than *what* is done. A study of the way the world behaves makes it possible to determine what functional relationships exist between variables. A judicious use of simulation then makes it possible to determine how functional relationships may be changed. Rather than enshrining the conventional wisdom, models may make it possible to examine a range of untried relationships.

6. *The model as a data source.* The idea that a model can be used as a data source to compare systems that do not exist is basic to the systems approach. The model is used as a surrogate for unbuilt future systems and becomes the most reliable source of planning information. Models are essential when development of the system is an irreversible process, and the use of the model as a vehicle for simulation to generate data, which then can be analyzed in conventional ways, is one of the most important spin-offs of the systems analysis methodology.

THE BENEFITS

Given the problems of this methodology and the difficulties to be encountered in implementing it, what are the benefits? First, the method provides a mechanism for dealing with complexity. As Ashby (1956, p. 5) has pointed out, science traditionally has remained limited to investigations of things which are either intrinsically simple or which can be made so by assumption. These procedures provide a method for dealing with complexity directly.

Second, our methods accord with common sense. Human activity is more a matter of adapting to the vicissitudes of life than making predictions which can then be followed with assurance.

Third, the methods provide a way of analysing time-series data which is independent of the usual assumptions of independence and linearity.

Finally, the methods provide a basis for simulation to assess the effects of innovation prior to taking action in the real world.

The problems of managing health systems can no longer be dealt with piecemeal by components within the system suboptimizing without regard to the performance of the overall system. Patient care, for example, cannot be understood or explained by looking just at nurse–patient interaction. Integration and coordination can only be acquired by looking at system components in interaction.

Although systems analysis is necessary to the solution of health care problems, methods for doing it leave much to be desired. I believe the major sources of difficulty is the economic rationale and the way systems have been partitioned for suboptimization. Economic models are appropriate for resource allocation decisions but not for resource utilization. These problems cannot be studied in isolation. Cybernetic models may be used to describe the adaptive behavior of system components engaged in resource utilization. Once this process is understood, it may be possible to develop the trade-off relationships that can be used for economic analysis.

REFERENCES

Ashby, W. R. *An Introduction to Cybernetics.* New York: Wiley, 1956.

Ashby, W. R. *Design for a Brain.* (2nd ed.) New York: Wiley, 1960.

Bernard, C. *Introduction à l'étude de la médicine expérimentale.* Paris: Bailliere, 1865.

Cannon, W. B. *The Wisdom of the Body.* New York: W. W. Norton, 1932.

Enthoven, A. Systems analysis and the Navy. In F. Uhlig (ed.), *Naval Review 1965.* Annapolis: U.S. Naval Institute, 1964.

Hitch, C. J., and McKean, R. N. *The Economics of Defense in the Nuclear Age.* Cambridge, Mass.: Harvard University Press, 1960.

Howland, D. A. A model for hospital system planning. In G. Kreweas and G. Moriat (eds.), *Proceedings of the Third International Conference on Operating Research.* Paris: Dunod, 1964.

Howland, D. Toward a community health system model. In A. Sheldon, S. Baker, and C. P. McLaughlin (eds.), *Systems and Medical Care.* Cambridge, Mass.: M.I.T. Press, 1970.

Howland, D., and McDowell, W. The measurement of patient care: A conceptual framework. *Nursing Research*, 1064, *13*, 4-7.

LaRue, R. Mediplot. Unpublished working paper, Ohio State University, 1969.

Reid, R. A. An Evaluation of a Methodology for the Analysis of Time Series

Behavioral Data. Unpublished doctoral dissertation. Ohio State University, 1970.

U.S. Congress, Senate. Amendments to Title IX of the Public Health Service Act, Public Law 91-515, 91st Cong., 2nd. Sess., 1970, H. R. 17570, 84 Stat. 101-602.

U.S. Congress Subcommittee on Economy in Government, Joint Economic Subcommittee. *The Analysis and Evaluation of Public Expenditures: The PPB system.* Washington, D.C.: United States Government Printing Office, 1969, 3 vols.

U.S. Senate Committee on Government Operations. *Planning, Programming, Budgeting.* Washington, D.C.: United States Government Printing Office, 1970.

Wiener, N. *Cybernetics.* New York: Wiley, 1948.

CHAPTER 10

Observation and Measurement of Health Systems Behavior

Lillian M. Pierce

Whenever the subject of a discussion is observation and measurement, certain questions are almost sure to be asked: For what purpose are the observations and/or measurements being made? What is the nature of the phenomenon being observed or measured? What is the source of the data, the population, and the sample? What instruments are being used? Are the instruments valid? Are they reliable? What about the problem of bias? These questions are appropriate whether the phenomenon of interest is systems behavior or some well-defined attribute about which data can be collected with standardized instruments.

In response to the first two questions, the purpose of our observations was to determine the nature of the phenomenon, that is, to describe system behavior in terms of the patient's response to the allocation of hospital resources.

The source of the data was the system itself; the interaction of the patient and various members of the health team. The population, the sample, will be described later. There were no standardized instruments for obtaining the kind of data we needed. The methods we used will also be described later in this paper, along with problems of validity, reliability, and bias.

I shall discuss specifically the methods we used to collect and reduce the data for the nurse-monitor in the patient-care system study, described in Howland, Pierce, and Gardner (1970). As Ashby (1956) says, "Every system, fundamentally, is investigated by the collection of a long protocol, drawn out in time, showing the sequence of input and output states" (p. 88). That, in

123

essence, is what our data consist of — a very long protocol drawn out over an extended period of time, which does, in fact, show the sequence of input and output states.

To develop a model that accurately describes the behavior of the system in terms of the patient's response to the allocation of hospital resources, we needed comprehensive data that describe the interaction of the patient with all the various members of the health team and in the variety of settings throughout the hospital. Since we view the system as adaptive, it is not possible to predict, a priori, when or where interactions or changes will occur.

To obtain the comprehensive data we believed were necessary to develop the model, we decided to observe patients continuously, 24 hours a day, from admission through discharge. We realized that this would be a monumental task and explored a number of different methods of collecting the data.

During the earlier work on this phase of the project some of us had collected data in the operating room by visual observation and manual recording of events as they occurred. The manual recording was supplemented by a tape recording made by the anesthesiologist, who wore a small field microphone and provided a running commentary of his observations of the patient's state and his own activities. We collected data for 18 different patients during this phase. The time span for the observations was relatively short, particularly when compared with observations in the nursing unit; the number of variables to be considered were relatively few; and the situation was controlled by the anesthesiologist and the surgeon. The data collected were then used to begin developing the lists of patient state and regulatory variables and the first rough plots.

In developing our plans to collect data in other settings, we questioned whether the same techniques could be used. We were particularly concerned about the length of time of the observations and the amount of data to be recorded. We considered using electronic recording equipment but found this to be impractical when the patient was ambulatory. Available equipment was also limited in terms of the number and kind of variables that could be recorded. We also considered video recording. This is an excellent method for collecting continuous data, particularly in well-controlled settings such as the operating room, the recovery room, and the intensive care unit, where the patient is not ambulatory or where the time span for observations is relatively short — four to eight hours. The cameras can be ceiling-mounted and can be operated from a remote control room. Video recording has the added advantages of capturing the data in their entirety and total recall for the investigator as often as he wishes. Of course, the data must be extracted from the tape once it has been recorded. However, the high cost of the equipment and technical assistance and the low reliability of the equipment when it is used over very long periods of time, as well as the limited range of the camera's scope when the patient is ambulatory, made video recording impractical for our purposes. After much deliberation, we decided that nonparticipant observation with manual recording of the data in the form of a narrative description of events as they occurred would give us what we needed. Identifying the kind of data we needed and the procedures we intended to use was the first step.

During the early phase of the study, when data were being collected in the operating room, patients requiring different kinds of surgical procedures were observed. When the study was extended to include other settings in the hospital we decided to observe only patients who were having abdominal hysterectomies. We had several reasons for selecting these particular patients. We elected surgical over medical patients because of the greater uncertainty of diagnosis and of length of stay for medical patients. Of all the various kinds of surgical patients we could have observed, we selected abdominal hysterectomy patients because (1) sufficient operations were being done to give us a substantial population from which to draw data; (2) the patients, generally, had no pathological condition other than that for which the surgery was indicated; (3) the surgical procedure is relatively straightforward and the length of time involved is generally consistent over patients; (4) patients had homogeneity, at least of sex; and (5) age range was consistent, from 35 to 50 years. There was some question about the psychological problems of women undergoing abdominal hysterectomy. However, after a rather extensive review of the literature we drew certain conclusions. First, the psychological problems of gynecological patients, particularly those having hysterectomies, have been investigated more extensively than other categories of surgical patients; and second, there is no general agreement about the problems these patients may have. The nature of the study did not demand a homogeneous sample beyond the rather general characteristics described.

The usual procedure at Ohio State University Hospitals is for the patient to be admitted between 2:00 P.M. and 4:00 P.M. on the afternoon prior to the day on which surgery is scheduled. The surgical gynecological staff indicated the average length of stay for abdominal hysterectomy patients as 7 to 10 days. Anticipating a 10-day stay as maximum, we were faced with the problem of observing and recording data continuously for approximately 240 hours for each patient we observed. We discovered later that the staff had underestimated the length of stay.

We decided to use nurses to collect the data, again for several reasons. First, observation is considered to be an essential part of the nursing process. Second, the hospital would be a familiar environment and nurses would be less concerned with their surroundings than would observers from another discipline. Finally, nurses would be able to identify various personnel interacting with the patient and procedures being carried out (Pierce, 1970; Howland, Pierce, and Gardener, 1970).

The very fact that observation is an essential part of the nursing process can be a possible source of observer bias when nurses are used to collect data in the patient-care setting. Since nurses do make inferences based on their observations, the recordings may be the observer's inferences rather than objective data. Additionally, nurses with different preparation and different experiences may have a greater degree of difference in observational skills than can be attributed to individual differences. To counteract the possibility of observer bias and different degrees of observational skills, our data collection procedures were designed with particular attention to detail and the observers trained accordingly. We tested a variety of techniques, examined the data collected in terms of the needs of the study as defined by the conceptual framework, rejected the techniques that did not provide adequate data, and developed different ones until we were satisfied. We then prepared a Data Collection Manual (Pierce, Brunner, and Larabee, 1968a), essentially a manual of operating procedures. The manual was designed to orient the observers to the project, indicate what was to be observed, and explain how the observations were to be recorded. The function of the data collection was to observe and record, as objectively as possible, "who is doing what for the patient and how the patient responds" or "what is happening with the patient and who is doing what about it." No judgments and no evaluations were sought.

All the original observers participated in an intensive training program in which we used a variety of teaching methods, beginning with a 30-minute videotape of patient interaction with a variety of personnel. The tape was especially prepared to demonstrate nonparticipant observation techniques, but it proved to be an excellent practice device. When the observers were familiar with the techniques and demonstrated a consistent level of recording ability, they went in to a variety of hospital settings for practice sessions. The initial practice sessions were one hour long; these were gradually increased to four hours. As the sessions increased in length, we used a second observer in the same setting to check the adequacy of the observations and recordings of the first observer. We used a videotape recording and dual observation to determine inter-observer reliability, which was consistently .90 and above. We also used videotape recordings to determine intra-observer reliability over time; this, too, was consistently .90 and above. Having developed the data-collection procedures and trained the observers, we had taken our second step.

The observers were formed into six-member teams to collect the data. Each member of the team was responsible for the same four-hour block of time each day during the patient's hospital stay. Observations began when the patient was admitted to the nursing unit and continued until discharge. Wherever the patient went, the observer also went and recorded — again as objectively as possible — a narrative description of who did what to, for, or with the patient, and how the patient responded. A typical sequence of entries in the data log (see Figure 1) might read:

0858   Patient sitting on side of bed. Lab. tech. enters. Pt. says "I've been expecting this." Smiles, lies down. Lab. tech. asks pt. to extend right arm. Pt. extends right arm.

0859   Lab. tech. examines antecubital space, right, applies tourniquet, cleanses area with alcohol sponge.

0900   Removes tourniquet, obtains blood specimen, applies sponge; tech. instructs pt. to bend elbow. Pt. bends elbow, sits up, and smiles. Pt. states "that wasn't too bad."

0903   Lab. tech. exits room. Pt. holding sponge on VP site.

0905   M.D. enters pt's. room — greets patient. Pt. states "she is feeling apprehensive about surgery today." M.D. explains patient's surgery to her, checks pulse, and assures patient "everything is fine."

0908.   M.D. exits. Pt. calls home on telephone; starts conversation with husband.

0910   R.N. enters with hypodermic syringe in hand. Pt. stops telephone conversation. R.N. says, "Mrs. Brown, I'm going to give you an injection — you'll be going to surgery soon. Please lie down and turn on your right side."

0911   Pt. lies on right side. R.N. raises pt's gown, cleanses IM site with alcohol sponge, gives Demoral 100 mg & atropine 0.4 mg IM in left buttock. Lowers gown.

0912   R.N. tells pt. "I want to check your blood pressure." Pt. moves on back. R.N. applies B/P cuff, left arm.

0913   R.N. measures B/P, removes cuff, records B/P 128/82. Pt. closes eyes.

0915   R.N. exits room.

In addition to the narrative recordings in the data log, the observers also recorded data about the patient's physical and social environment, such socioeconomic factors as race, religion, marital status, and occupation; and medical parameters such as the medical history of the patient and her family.

Ten patients were observed. The length of stay ranged from nine to 24 days. With observations being made on a 24-hour basis, we collected approximately 3,000 hours of time-series data describing system behavior. Five complete sets of the data are presently being used for analysis.

As the data were being collected, we developed the procedures necessary to translate the verbal data to numbers on punch cards for computer analysis. The data were in the form of narrative description; "at this time, in this location, this person is doing this to, for, or with the patient with this result" or "at this time, in this location, the patient's state is this, with this being done by this person" or simply "at this time, in this location, the patient is doing this." An amazing amount of narration can be recorded in 24 hours, especially during the first two postoperative days. The reduction of the data

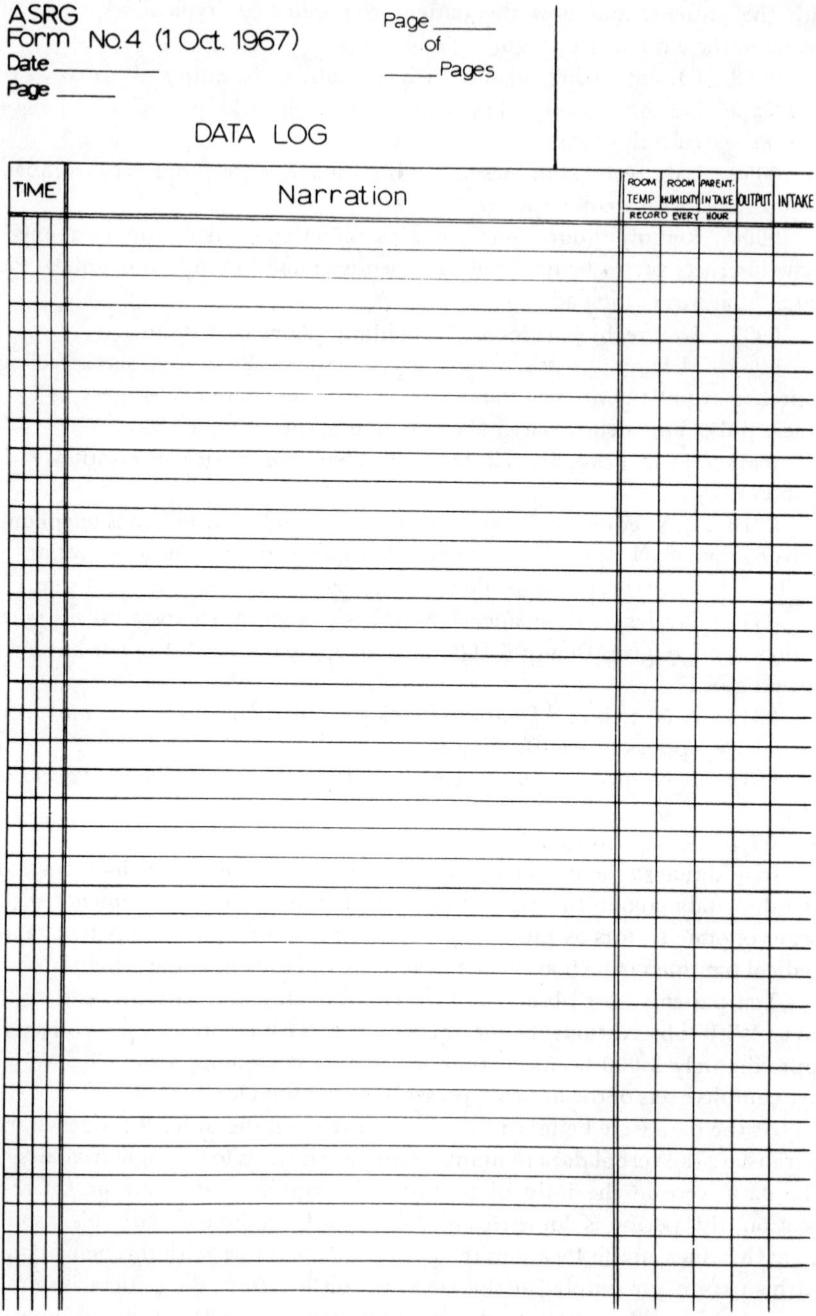

ASRG
Form No. 4 (1 Oct. 1967)
Date_____
Page _____

Page_____
of
_____ Pages

DATA LOG

TIME	Narration	ROOM TEMP	ROOM HUMIDITY	PARENT. INTAKE	OUTPUT	INTAKE
		RECORD EVERY HOUR				

Figure 1. Data log.

was a transitional process; basically a content analysis with the major categories, defined by the conceptual framework, as patient-state variables and resource (or regulatory) variables. Time, location, and participant were additional categories.

Although the conceptual framework provided the guidelines in terms of the major categories, the coding scheme — subcategories, inventory of variables, and the code evolved from the data. The two major problems were the large amount of data and the nature of the raw data. We were dealing with both qualitative and quantitative data. The quantitative data represented all the scales from nominal to ratio. This is a common problem.

Again, as with the data-collection procedures, the development of the data-reduction procedures involved trying out a technique to code parts of the recorded data, subjecting the coded data to cursory analysis, and revising the procedure when necessary. When the procedures were developed to the satisfaction of the entire team we prepared another operating manual, the Data Reduction Manual (Pierce, Brunner, and Larabee, 1968b). The manual was designed to orient the observers to the coding process, provide instructions and guidelines for coding, and, in addition, provide the content necessary for coding (word to number). The word-to-number content of the manual was expanded as additional data were coded. The observers were prepared as intensively to code the data as they were to collect them. When they had developed a high level of skill and inter-coder reliability had been established, two-member coding teams were formed. Collection and coding activities both took place concurrently. When not collecting data, observers coded that collected earlier, but not recorded by themselves. A positive byproduct of the simultaneous collection and coding activities turned out to be an increase in the quality of the recordings — both the content and the handwriting. The narrations became more concise, more specific.

Figure 2 is a sample of the data coding sheet we used to reduce the data for analysis. The coding sheet is divided, top to bottom, into three major sections. The top section is used primarily for coding identification and record-keeping purposes. The second section is used for data control and patient identification. Each small block represents one column on the keypunch card. The numbers at the top are the numbers of the columns. The title headings identify a field. The third section of the coding sheet is used to enter the appropriate code numbers representing the narrative data. This section is divided into 13 fields. Each field contains a specific category of data. Precise instructions for entering code numbers are contained in the coding manual. Each horizontal line represents one keypunch card. All entries in the second section of the coding sheet are repeated by the keypunch operator for each line on the coding sheet. Columns 1 through 19 will contain the same data on each card. Columns 20 through 79 will contain different data on each card based on the entries made in each of the fields during the reduction process. The fields can be identified by the dark vertical lines on the coding sheet. Some of the major fields have subheadings. For example, the field covered by columns 30-34 has a title heading *Activity* and subheadings *Verbs* and *Desc* (descriptive terms).

The coding manual was set up in sections with each section representing a field in the code format. Each section contains specific instructions for entering code numbers and the word-number content. Moving from left to right across the code sheet the first three fields (columns 20-23) are labeled *Time*, an essential bit of data that must be entered on every card; *C* which identifies the card number in the sequence; and *Loc* the location where the observation was made. The next four fields, columns 27-52, represent the major category of resource (or regulatory) variables. *Pers* (columns 27-29) identifies the personnel involved. *Activity* (columns 30-34) has subheadings *Verb* and *Desc*. Numbers entered in this field indicate the activity being carried out by the personnel involved. The use of subfields is designed for conservation purposes. When we first began to develop the inventory of variables we found ourselves with a list of activities of unbelievable length. We found the same verbs occurring over and over again; for example, change bed, change dressing, change IV bottle. We also found the same object occurring repeatedly; change dressing, reinforce dressing, check dressing, adjust dressing. By developing a list of verbs and a list of descriptive terms which could be used interchangeably, we markedly reduced our list of activities. We did the same thing for the list of patient activities (columns 67-72). Columns 35-37 and 73-75 are both anatomical indices, and both use the same code content: columns 35-37 in relation to personnel, columns 73-75 in relation to the patient. Columns 38-52 are used when the patient is given a medication to identify the agent, dosage, the unit, and the route of administration.

The next two fields (columns 53-56, 57-59) are used to indicate the occurence and receiver of a verbal message. The sender of the message is identified in the personnel column, even if the sender is the patient. We have not attempted to categorize verbal messages. As verbal messages occur they are numbered in sequence and the sequential number becomes the code number.

The next three fields (columns 60-75) represent the major category of patient-state variables. The first field *Pat. State* will contain entries of physiological variables and their values; for example, systolic blood pressure and the appropriate value or pulse and the rate of temperature and the value. The next field *Pat. Activity* will contain entries that identify physical activities of the patient — walking, standing, sitting, reading, lying, and so on. Once again, the activities are coded by using various combinations of verbs and descriptive terms. The *Anatomical Index* entry here is used in reference to patient-state or activity variables. The last field (columns 76-79) is used to

DATA CODING SHEET

PROJECT IDENTIFICATION	DATE CODED	CODED BY	PUNCHED BY	VERIFIED BY	RAW DATA PAGE	CODE SHEET PAGE	OF
RF - 13846	5/15/68	D.S. & S.W			19	54	

PATIENT IDENTIFICATION (0: 1–9)	ROOM Nº (1: 0–2)	HOSP. DAY (1: 3–4)	POST DAY (1: 5–6)	C.S. PAGE (1: 7–9)
900790747	923	05	01	051

A. COL. 20-26 MUST BE FILLED ON ANY LINE USED.
B. EXCEPT FOR COL. 38-52, ANY FIELD USED MUST BE COMPLETELY FILLED.
C USE OF "ENTER" VERBS REQUIRES THE USE OF "EXIT" VERBS AT APPROPRIATE TIMES.
D WHEN CODING VERBAL MESSAGES COL. 27-29 AND 53-59 MUST BE FILLED.
E. AT TIME 0001 START A NEW C.S. AND INCREMENT COL. 13-14 AND IF APPROPRIATE, 15-16.
F. WHEN AN OBS. OR MSMT. IS CODED COL. 60-66 MUST BE FILLED
G. WHEN COL. 60-66 ARE USED, AN OBS. OR MSMT. SHOULD BE IN COL. 30-34 IF APPROPRIATE

TIME	C.	LOC	PERS.	ACTIVITY VERB	ACTIVITY DESC.	ANAT. INDEX	PHARMACEUTICAL AGENT	DOSAGE	U.	R.	VERBAL MESSAGE	REC.	PAT. STATE KIND	PAT. STATE VALUE	PAT. ACTIVITY VERB	PAT. ACTIVITY DESC	ANAT. INDEX	ENVIRON-MENT
0954		01	241	27	002								002	0080	149	804		
0954		01	201	27	072								073	0025				
0954		01	201	27	087								078	1032				
0959		01	201	99	000										144	739		
0959		01	010								0188	990			126	750		
1000		01													144	753		3000
1000		01		99	000										144	700		
1000		01	201	98	000													
1000		01	201	26	200													

RF 1651, FORM #12 (20 JUNE 1968)

Figure 2. Data coding sheet.

enter data about the environment — temperature, humidity, light, and noise.

Because we were also concerned with the physician's orders for the patient, we developed a similar coding scheme to handle these data. The code content and specific instructions are also contained in the manual.

During the coding process we did not work directly from the Data Reduction Manual. Once the teams were familiar with the various fields and their content, and with specifics of the coding process, we set up a Kardex as a working device. The word-number content for each category of data occupies from one to nine cards.

Collecting and coding the data was the third step; we are then ready for the analysis.

During the course of this project we have been asked a number of questions about our data-collection and data-reduction procedures, particularly in respect to the presence of an observer. How does the patient respond? Does it bother her? We interviewed each of the patients when they returned to the clinic or to their respective physician in an attempt to learn their reactions to having been observed. To the best of our knowledge patients were not bothered. I might add that when we asked a patient to participate in the study, we told her that an observer would be with her at all times, would go wherever she went, but would not do anything for her. Each patient was also assured that the observations would be discontinued if the observer's presence became annoying. None of the patients asked to have observations discontinued.

We have also been asked about the effect of the observer on nursing and other personnel. Does it change system behavior? Undoubtedly it does, to some extent. However, we can say with a relatively high degree of confidence that personnel seem to behave in much the same way whether an observer is present or not, particularly after they become used to the observer being there. All personnel and patients were aware that the observers were recording what happened and were not evaluating personnel performance. The narrative data describe patient states that require response with no response forthcoming: the data also describe personnel behaviors which do not always seem appropriate.

When we developed the procedures for collecting the data, we were concerned about the length of time, amount of data to be recorded, and number of events that might be occurring at the same time or in such rapid sucession the observers would be unable to record them. We had very few problems with time or amount of data. Because our data tend to be macroscopic rather than microscopic, and because we did develop the procedures in considerable detail, the observers had very little difficulty collecting data adequate to our purpose. We did have two problems with the observers. There were some who, when observing the patient, could not tolerate identifying some action which seemed to be indicated and then not doing something about it. The second problem was boredom during the periods of little or no interaction. The observers described those periods as more exhausting than when there was considerable activity to be recorded.

REFERENCES

Ashby, W. R. *An Introduction to Cybernetics.* New York: Wiley, 1956.

Howland, D., Pierce, L., and Gardner, M. The Nurse-Monitor in a Patient-Care System. Unpublished final report of Grant No. NU00095, Adaptive Systems Research Group, Ohio State University, 1970.

Pierce, L. M. Development of a Patient-Care Model. Paper presented at the Annual Meeting of the Division of Nursing, USPHS, Department of Health, Education, and Welfare, Bethesda, Maryland, March, 1970.

Pierce, L., Brunner, N., and Larabee, J. Data Collection Manual. Unpublished Paper, Ohio State University, Adaptive Systems Research Group, Columbus, Ohio, 1968. (a)

Pierce, L., Brunner, N., and Larabee, J. Data Reduction Manual. Unpublished Paper, Ohio State University, Adaptive Systems Research Group, Columbus, Ohio, 1968. (b)

CHAPTER 11

Computer Analysis of Time-Series Data

Robert D. LaRue

Lord Kelvin once said, "When you can express what you are talking about in numbers, then you know something about it. . . ." Researchers are now able to transform information into numbers. The objective of this paper is to discuss various aspects of using the computer for analysis of masses of numbers.

Unfortunately, many investigators who could benefit greatly from a computing system are not taking advantage of this research tool. In some cases, they lack understanding of the computer and are intimidated by some of the popular misconceptions about it.

First, the computer is not a highly intelligent superbrain. While some of the results obtained from computer usage may seem somewhat miraculous, they are obtained because the computer is capable of carrying out simple computations at a tremendously high rate of speed and because a human programmer was able to tell the computer how to combine these basic operations into a sequence that produced the required output.

The computer is reputed to be an extremely accurate device. This is not always true. We know that adding 1,000 ones together and dividing by ten should produce the same numerical results as adding 1,000 tenths together. On a test, one computer gave 100.00 as the result of the first series of operations, but 99.82 for the second series of operations.

Some computers behave like obstreperous children. For example, one of my students produced a program from which he was obtaining incorrect results. The errors could easily be eliminated by inserting the statement:

INTEGER APRIL

into his program. When the student returned on the following day he had exactly the same output as on the previous run. However, the statement he had inserted was

INTIGER APRIL

The message from the computer was:

INTEGER MISSPELLED *** WARNING ONLY

Here it appears that the computer knew exactly what the student was attempting to do but was not going to help him at all. Results of this type can be frustrating.

Another misconception regarding computer use is that one must be a proficient mathematician to understand and write programs. While there could be some truth in this statement in the area of problem-solving programs, it certainly is not true in the area of data processing. Anyone capable of carrying out a research investigation should be able to understand the fundamentals of programming.

Most researchers will have neither the time nor the inclination to become expert programmers. However, in this case the researcher and programmer must work as a team, and this effort will generally be much more productive if the researcher takes the time to learn elements of programming. It is much easier for him to obtain this knowledge than for the programmer to become equally proficient in the researcher's field.

Once a program has been written we are confronted with the task of inputting data into the computing system. Approximately 10,000 cards were required to contain information collected on each of the patients in our study. [The study was carried out at The Ohio State University by Daniel Howland and co-workers. Ed.] Even with a fairly high speed card reader (1,000 cards per minute), the computer needs considerable time, relatively speaking, for the input phase before the actual analysis can be done. Another option is to input data directly onto magnetic tape (if the proper device is available) or can be transferred from cards to tape using one of the many computer utility programs available. High-density magnetic tape has the capacity for storing more than 200,000 cards of data on one 2,400-foot roll. This is the equivalent of data for 20 patients. It is wise either to input data directly onto tape or to transfer cards to tape, for, in addition to saving computer storage, it saves the inputting of data every time one wishes to do an analysis; it also saves storage.

There are two general modes in which data may be processed. Batch or "hands off" mode is characteristic of many computing systems. In this mode of operation, the user prepares a program, submits it to a system terminal, and returns at some later time to obtain the output from the job. Turn-around time (TAT)— the amount of time between submission and return of a job — depends upon the capacity of the system, size of the job, and the current load on the system. At Ohio State University, which has several large systems, TAT may be less than five minutes on small jobs with light user loads or longer than 24 hours when medium or large jobs are processed with high user loads. Long TATs seem to reduce the effectiveness of the research process, for the investigator may, in the interim, have lost his trend of thinking about the problem and will then require added time to become again immersed in his data.

The interactive or "hands on" mode of operation can be a most effective way of using a computing system. This mode may involve time sharing a large computer with other users (each user has his own typewriter terminal), or operating a small computing system during the data processing process.

In time-share mode, the user may notice some relatively short delays during the operation. However, for the most part, he feels as if he were the only user of the system. It has been found in problem-solving courses that students using the time-share mode develop an insight into the fundamentals involved much more rapidly than when using the more traditional approach. If the results obtained are incorrect, immediate modification of the program and a rerun will enable the individual to determine whether he has located the errors of his original approach. This approach seems to have a great deal of promise for the researcher engaged in systems oriented research. He will have the capability of stopping a program during execution, modifying parameters, and restarting the program. Results obtained by this mode of operation would require complex programming or a considerable number of runs over a long period of time in batch mode. Also, the insight developed by the user cannot be overlooked as a factor in making the research more effective.

One of Confucius' many sayings is that "a picture is better than a thousand words." This statement can be applied to the representation of numerical information as well as to descriptive phrases. A picture of a hummingbird was produced by a computer controlled plotter. The plotter moved from one to another of thousands of data points stored as numbers in the computer memory. A display of the numerical values involved would require several pages of printer output and would certainly not have the significance to the viewer that the drawing would.

While the plotter produces exact graphical output, graphical approximations can be obtained from a printer or typewriter. A bar graph is a typical example of a graphical output. A similar display can approximate a curve.

CHAPTER 12

Nurse-Patient-Physician Communicative Interaction Process

M. Jean Daubenmire

Anyone who is familiar with hospitals or health care systems, either as a professional or as a patient, is aware of the frequent communication that says verbally "How are you this morning" but nonverbally "Please don't tell me how you are." For example, the nurse with a tray of medications in hand edging toward the door as the patient is trying desperately to tell her how she is, or the physicians who enter the patient's room en masse and ask how the patient is but continue to converse with each other about the pathophysiology involved.

Communication is regarded as a key concept of all social organization and of the behavior of all systems (Abdellah, 1966; Cherry, 1966; Miller, 1965; Ruesch, 1959). Medical and nursing literature also recognizes the importance of communication. Smith (1964) says that "the science of communication is more pertinent to nursing than the science of disease or pathology" (p. 70). Although there is general recognition of the importance of communication, there is still much miscommunication or lack of communication in the everyday world of the hospital.

In reporting this study, I am attempting to develop a methodology to examine the nature of the nurse-patient-physician communicative interaction system within a conceptual frame of reference. Many factors in both nursing and communication research have led me to use a systems approach rather than one of the more classical methodologies. My aim is to identify patterns of interaction and determine their effects on the hospitalized patient. The method is descriptive rather than explanatory; that is, it is concerned with *how* rather than *why* the interaction system operates.

139

This study assumes that the patient enters the hospital in a state of disequilibrium, psychological as well as physiological. Psychological disequilibrium is thought to be due in large measure to: (1) separation from familiar surroundings; (2) interruption of daily living activities; (3) fear of the unknown; and (4) fear of undergoing diagnostic tests, therapy, anesthesia, or surgery. An understanding of the nurse-patient-physician communicative interaction process must be based on a knowledge of the way a patient moves from the community, through the health care system, and hopefully back into the community. This is illustrated in the model of adaptation and health in Figure 1.

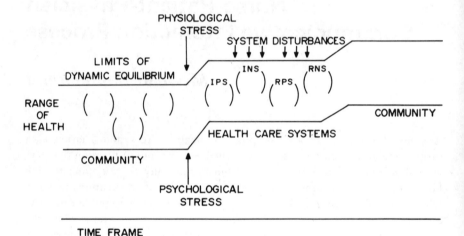

Figure 1.   Adaptation and health.

(INS = initial   nurse   state,   IPS = initial   patient   state,
RPS = resultant patient state, RNS = resultant nurse state. Brackets
are vectors describing the state of the individual including the values
of the essential variables at a given point in time.)

One of the most important ways that persons adapt to or change their environment is through communication. In the home or community, an individual develops effective ways of communicating with persons or objects in the environment which allow him to stay within some limits of dynamic equilibrium. When an individual enters a health care system, he enters a new and different environment, an environment in which the focus is on system-determined requirements for behaviors of all participants, i.e., rules and procedures established for the organization, rather than the individual patient. The patient is faced with the need to adapt by finding ways of communicating with people in the environment. Yet, it is at this time that the patient experiences stress and has fewer personal resources available for adapting. Bartter and Delea (1962) report that the impact of entering and adjusting to the hospital may affect the adaptive processes of a patient for as long as several days after admission. Pride (1968) indicates that "nursing activity affects the adaptive processes of patients; and the adaptive processes of patients are enhanced by nursing care planned to meet individual patient needs" (p. 293).

Wiens, Thompson, Matarazzo, J. D., Matarazzo, R. G., and Saslow (1965) point out that "some functions of the professional nurse can be carried out *only* through effective verbal and nonverbal communication, for example, teaching health practices, extending emotional support, bringing about changes in patient behavior, and obtaining an adequate health history to evaluate the patient's present health status. It follows then that the nurse's communication patterns should be related to her success in a nursing role" (p. 322). King (1968) defines nursing as "a process of action, reaction, interaction and transaction" (p. 27). She indicates that nurses perform their functions within social institutions and in interaction with patients and personnel. All these factors have bearing on the essential communication the nurse seeks to establish.

The hospitalized patient is dependent primarily on medical and nursing personnel for information and frequently for the carrying out of his daily activities. Much of the uniquely therapeutic communication that occurs between nurse and patient appears to be initiated intuitively. Nurses themselves find it difficult to identify what types of communicative relationships seem effective in helping the patient and have difficulty even in repeating the process. Therefore, nurses might benefit from a better understanding of the communicative-interaction process so that its components can be identified, repeated, taught, and made standard in nursing practice.

The study is based on a concept of adaptive systems which describes complex systems and their adapative behavior, and on the use of certain principles from communication theory to identify factors in the communication process.

While it is easy to agree in essence with the concepts of general systems theory, it is another matter to apply these concepts in nursing research. However, general factors justify the use of this theory to help obtain answers to the questions we are asking about the practice of nursing.

The first of these factors is the need to view the nurse, patient, and physician as a system. A number of studies have been conducted concerning the communication interaction process of nurse-patient, physician-patient, or nurse-physician. However, no studies have been found which deal with the communication process of nurse-patient-physician. One could assume that each is affected by the other and by the system within which they all operate. While my primary interest is in the nurse, patient, and physician, my data include all patient-family-health team interactions.

When the nurse-patient-physician situation is viewed as a system, then, a problem previously identified as a patient's problem may also be a problem of the nurse or the system. This perspective requires a different method of intervention in patient care. For example, if the nurse working with a patient who is demonstrating a great deal of hostility plans her intervention by analyzing factors within the system, she may view the hostility as being this patient's only alternative reaction to a very difficult situation, rather than labeling him a "difficult patient."

A second factor emphasizing the need for systems theory in this field is the potential increased use of existing analysis schemes. Because of the complexity of the nurse-patient-physician interaction, any predetermined observation schedules or analysis schemes demand that all essential variables affecting multiperson interacting systems also be predetermined. In other disciplines, such as psychology and sociology, sufficient descriptive data have been accumulated to meet this demand. This is not the case in nursing. Much more descriptive data in nursing must be gathered to acquire the knowledge necessary to identify essential variables.

The fact that most studies have been limited to verbal interaction is a third justification for the use of systems theory. Nearly all studies of nurse-patient interaction have been confined to the study of verbal interaction for which various categorization and analysis schemes have been devised (Diers and Leonard, 1966; Johnson, 1964; Meyers, 1964). Although these studies may add to an understanding of types of verbal communication, it is unlikely they can contribute to understanding the overall process since as much, if not more, of one's communication with other individuals occurs at a nonverbal level. Communication theorists attest to the impossibility of separating the verbal and nonverbal behavior (Cherry, 1966; Ruesch, 1959; Watzlawick, Beavin, and Jackson, 1967).

Finally, systems theory would aid in considering the idea of content versus process in communication. Many authors refer to verbal messages as the content of communication and to nonverbal behavior as the process. Diers and Leonard (1966) state in their study of communication, "two kinds of dimensions that are worth keeping distinct are *content* and *process*, i.e., semantic (what is said) and syntactic (how it is said)" (p. 226).

I believe that it is not useful to separate these, because what is said (verbal) and how it is said (nonverbal) are both parts of the message. Further, how it is said is not a description of the process. An understanding of the process can only be identified as the *effect* of the message (both verbal and nonverbal) upon the changing state of the participants which can be observed.

Recognition of the limitations of the experimental method also encourages the application of systems theory to this field. Nursing literature reports numerous studies using experimental methods. The investigators have attempted to measure the effect on a patient of a specific approach by one or more nurses (Chapman, 1970; Kaufmann, 1964; Pride, 1968). The experimental method requires that situational variables be controlled and like samples selected. Johnson, Johnson, and Dumas (1970) attest to the difficulty of controlling situational variables in nursing research: "There are a multitude of factors influencing the situation in which a nurse cares for a patient. The nursing care is not given in isolation from these factors. Neither can research in nursing care be conducted as if these factors had no influence on the research" (p. 341). Abdellah (1970), in a review of nursing research, also refers to the problem: "Since there are so many extraneous variables in the situation, both organismic and environmental, it is exceedingly difficult to keep the variables under sufficient control" (p. 10).

I contend that it is not useful to attempt to control these variables at the present time. Perhaps in nursing research one of our errors has been trying to find ways of making the nurse-patient-physician interaction process simple instead of searching for ways of studying and understanding its complexity. As Ashby (1956) has written:

Science today stands on something of a divide. For two centuries it has been exploring systems that are either instrinsically simple or that are capable of being analyzed into simple components. The fact that such a dogma as "vary the factors one at a time" could be accepted for a century, shows that scientists were largely concerned in investigating such systems as allowed this method; for this method is often fundamentally impossible in the complex system — they are so dynamic and interconnected that the alteration of one factor immediately acts as cause to evoke alterations in others, perhaps in a great many others (p. 5).

In summary, we need research methodologies that deal with both the individuality and the complexity of nursing situations, and systems methodologies seem to include these requirements.

Communication theories have developed from Shannon's and Weaver's (1949) early linear models, which described the source, encoder, message, channel, decoder, and receiver. The linear model was later modified by Wiener (1961) who added the cybernetic concept of feedback. Wiener's cybernetic model is demonstrated in Figure 2. In feedback, the result of the effector's (receiver's) activity is monitored back to the receptor (source) so that the system is self-regulating (von Bertalanffy, 1956). Individuals use feedback to alter or clarify messages. Later theorists have attempted to conceptualize the communication process as one of a helix model resembling a spiral that is moving forward but is dependent on the past, which influences the present and the future (Dance, 1967).

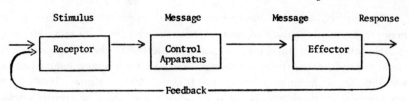

Figure 2.   General systems theory model of communication.
(Reprinted from Ludwig von Bertalanffy, "General System Theory,"
in *General Systems: Yearbook of the Society of General Systems,* Vol.
1, 1956, published by the Society for General Systems Research.)

All these models are helpful in conceptualizing the elements of communication. However, when one attempts to analyze communication in terms of source, message, and receiver one collects isolated bits of information, without the connecting link—the process. The content of a message can be determined, the identification of source or receiver can be noted, and the frequency of initiation or reception can be counted, but these facts give little information as to what happens to two or more persons in the process of communicative interaction.

Another trend in communications research has been the categorizing of messages as to requests, statements, content, orientation. Here again, this gives statistical information about the number of requests from a number of patients and about whether most of an individual's statements were oriented toward himself or others. But again it gives no insight as to how two or more people affect each other in a relationship through verbal and nonverbal communication.

The study of human communication can be divided into three areas; syntactics, semantics, and pragmatics (Morris, 1938). Syntactics refers to the transmission of information. The main concern of semantics is meaning. Pragmatics refers to the behavioral effects of communication. All three levels concern signs and rules about signs. The level that would seem most applicable to the study of the nurse-patient-physician communicative interaction process is pragmatics. Cherry (1966) refers to pragmatics as the "real life" level. Thus, from the perspective of pragmatics, all behavior, not only speech, is communication, and all communication—even the communication clues in an interpersonal context—affects behavior (Watzlawick et al., 1967). Shannon and Weaver (1949) have defined communication as "all those procedures by which one mind may affect the other" (p. 3). Ruesch (1967) further defines communication as follows:

A statement is made by a person or a group. The statement becomes a message only when it is followed by a reply or action resulting from the statement. After repeated exchange back and forth, the identity of the participants, the instructions for interpretation, and the properties of the symbols are clarified. It is possible that in the process of exchange, the intentions and anticipations of the participants may change (p. 60).

Birdwhistell (1952) states that an individual does not communicate but rather he engages in or becomes part of communication. "He does not originate communication, he participates in it. Communication as a system, then is not to be understood on a simple model of action and reaction, however complexly stated. As a system it is to be comprehended on the transactional level" (p. 104).

Thus, theorists describe communication as being very complex with many variables involved. There is an implied back-and-forth action, the possibility that the participants may change as a result of the process, and the importance of the effect of the communication on the participants rather than emphasis on the content of the message.

While studies in communication were not helpful in studying the problem at hand, concepts from communication theory were used in developing the overall approach to the study of the nurse-patient-physician communicative relationship. First, "people relate to each other through communication — a process which can be both observed and experienced. The intraorganismic functions of men which include such phenomena as feeling, thinking, and memory at present can only be postulated. Their exploration is dependent upon a two-person or multiperson situation in which the individual communicates his experiences or performs a task" (Ruesch, 1959, p. 897). Second, "the unit of study is not confined to a single individual but comprises all the people with whom a person habitually stands in communicative exchange" (Ruesch, 1959, p. 897). Third, one cannot *not* communicate; behavior has no opposite. "If it is accepted that all behavior in an interactional situation has message value, i.e., is communication, it follows that no matter how one may try, one cannot *not* communicate" (Watzlawick et al., 1967, p. 49). For example, activity or inactivity, words or silence all have message value; they influence others and these others, in turn, cannot *not* respond to these communications and are thus communicating. Fourth, "the word 'communicate' is generally understood to refer to nonverbal as well as verbal behavior within a social context. Thus, communication can mean interaction or transaction. Communication also includes all those symbols and clues used by persons giving and receiving meaning. Taken in this sense, the communication techniques which people use can be seen as reliable indicators of interpersonal functions" (Satir, 1968, p. 63). Fifth, pragmatics of communication deals with the present rather than the past. This approach recognizes that behavior is at least partly determined by past experience, but that the meaning of this experience must be inferred for each individual until it is communicated through verbal or nonverbal behavior.

This approach requires direct observation of communication between individuals with a "search for pattern in the here and now rather than for symbolic meaning, past causes or motivation" (Watzlawick et al., 1967, p. 45).

Nonverbal behavior has been a subject for investigation from the early 1900s, when it was studied as expressive behavior. More recent studies have been conducted within a framework of person-perception and emotion. Reviews of literature in this area have been done by Allport (1961) and Davitz (1964). Few studies have been conducted either with spontaneous behavior or within an ongoing interpersonal relationship. Therefore, the relevance of previous studies of nonverbal behavior to an ongoing nurse-patient-physician interaction with spontaneous behavior is somewhat limited.

Many studies of spontaneous interactive nonverbal behavior have, however, identified specific nonverbal behaviors that could be considered indicators of communication. These include body movement, eye contact, and facial expression. Dittman (1962) found patterns of body movement indicative of the patient's mood. Sainsbury (1955) found that the amount of movement was indicative of stress. Leventhal and Sharp (1965) described facial expressions as indices of comfort and indices of discomfort.

Several authors suggest that nonverbal behavior expresses the quality and changes of a relationship (Ekman, 1965). Ekman (1964), in a study of stress interviews, demonstrated that a very specific moment-to-moment relationship between verbal and nonverbal cues could be recognized by an observer. He identified that nonverbal behavior can serve a variety of communicative functions in relation to verbal behavior. Seven functions were described: repeating; contradicting; substituting for a verbal message; reflecting the person's feeling about his verbal statement; reflecting changes in the relationship; accenting parts of the verbal message; and maintaining the communicative flow.

In summary, much of the interpersonal research has been conducted in artificial groups, both social and problem oriented, and in interview or counseling sessions in an office setting. Very little research has been done on an ongoing process in a natural setting where the individuals are under stress. The nurse-patient-physician communicative interaction process is a dynamic ongoing process, it occurs in a natural setting and the individuals are under stress. Hence, what is needed is a way to describe the ongoing interpersonal process and to identify those essential variables relevant to the process and which are, to some extent, quantifiable. However, it is not enough to quantify these variables in terms of number and duration of interactions or in terms of who speaks to whom and about what. We must be concerned not only with the variations between individuals but also with the concomitant variation of psychological variables as measured by verbal and nonverbal behavior and physiological variables as measured by blood pressure, pulse, respiration, and so on for each individual.

Observation of variations of essential variables within and between individuals should lead to the identification of patterns of interaction and their effect on the hospitalized patient. Because the nurse takes action based upon the actual response of the individual patient and not upon the expected response of patients in general, we must have models that describe individual patient and nurse behavior. This requirement suggests the need for adaptive models rather than predictive system models.

We are defining the nurse-patient-physician communicative interaction system as one that proceeds by means of signs with which the nurse affects the behavior of the patient, the patient affects the behavior of the nurse, and both are affected by the system in which they operate. While this statement is generally accepted as a truism, we hope to provide some insight into the ways in which each participant affects the other and to identify environmental variables that enhance or deter the nursing process.

The nurse-patient-physician interaction process is a complex multiperson system with many interacting components and with many interacting variables within each component. The participants relate to each other through communicative behavior which can be observed and described. The primary function of communication is to enable participants to adapt to persons or objects in their environment, hence, the system is adaptive in nature.

Adaptive systems are open systems which respond to inputs from the environment, and therefore their behavior is dynamic, ongoing, and changing in time. They are goal directed or transactional in nature. Goals may be explicit or implicit. Implicit goals become explicit as they are communicated through verbal and nonverbal behavior. They are information limited. The nurse must have information to carry out her role, and the patient must have information to adapt to a new environment. They are resource limited. Howland and Colson (1970) state: "The overall behavior of systems is constrained by the capabilities and limitations of its component resources" (p. 3). Therefore, goal achievement must be considered in terms of the capabilities and limitations of both participants in the communicative interaction process. This process can be examined by identifying the observed effects of the message (both verbal and nonverbal) upon the changing states of the participants.

The focus of the study is the change in the patient's state or nurse's state over time as a result of communication. This is conceptualized in the time process model of human interaction (Figure 3). States of nurse and patient are described as vectors that include the values of the essential variables at a given moment in time. Feedback is represented by the arrows, which go forward in time. Feedback is usually conceived of as a means of altering or clarifying a message, and the model implies a return to the original state. The model regards either clarification or alteration of a previous message as a new

148

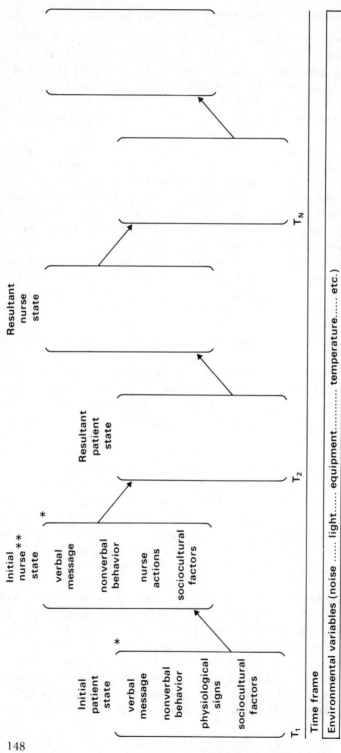

Figure 3. Time process model of human interaction.

(Reprinted from M. Jean Daubenmire and Imogene King. "Nursing Process Models: A Systems Approach." *Nursing Outlook*, 1973, *21*, (8), 514.)

message, because there is a different state for both patient and nurse. Therefore, each message is a new message — each state a new state in time — and the process can be observed as an alternating step function. While the task of integrating theories from widely divergent fields is an arduous one, it may well be possible to develop a theory of interaction based on an understanding of adaptive behaviors as they are communicated through verbal and nonverbal behaviors of persons in stress. Some basic requirements are that we shift our view from the individual patient to the processes that occur between people; we shift our focus of study from the individual patient, individual nurse, or physician to patient health team interactions; we look at the relationships between the transactions of various nurse-patient-physician systems and the patient's ability to stay within some limits of psychophysiological equilibrium.

A pilot study tested the feasibility of doing interaction research at varying times throughout a patient's hospitalization. Time-series data of patient-health team interaction were collected by means of videotape recording. Since the basic assumption of this study is that the patient enters a hospital in a state of psychophysiological disequilibrium, patients were selected simply on the basis of their and their physician's consent to participate in the study and were not screened by age, sex, or diagnosis. The study focused attention on the *process* by which information about the values of the various kinds of variables is communicated.

Because the communicative interaction process is a complex phenomenon that is transient in nature, the most useful method for collecting data is by videotape recording. The study required data that would provide both a microscopic and macroscopic view of the phenomenon: microscopic in terms of the behavior of the individual participant, macroscopic in terms of the reciprocal action of the participants within the environmental and social context. Videotaping also provides a means for keeping data intact so that it may be examined and reexamined as many times as necessary for identification of patterns of interaction. Observation and description of nonverbal data constitute the most difficult practical problem. Birdwhistell wrote extensively on the recording of nonverbal data, but the coding procedures developed by the Adaptive Systems Research Group at The Ohio State University are more useful for this study because they were developed in patient-care settings using the frame of reference of adaptive systems. In the current study we are in the process of expanding these procedures to allow for the coding of essential variables in the communicative interaction process using the guidelines in Figure 4.

At present we are looking at the behavior of nurse and patient microscopically to define descriptive measures of message modifiers such as, facial expression, tone of voice, gestures and body posture, so that these may be added to the vector describing patients and personnel.

1. Setting the scene: At the beginning of each transcription session, describe the setting in which the interactions are taking place. Include the where, who, and what is in the room. Include any equipment or machinery. Thereafter, only describe changes in the original setting.
2. Time, source, and receiver are recorded for all interactions.

Patient:

		Type of Recording
What he says	Verbal message	Exact content
How he says it	Nonverbal behavior defined as those behavioral mannerisms that accompany interactions between two or more persons	Facial expression Body position descriptors Scale of touch Verbal speech substitutes Patterns of movement as defined by spatial parameters of communication
Physiological	B.P.; Pulse; T.P.R.; Skin color, etc.	Only recorded when measured by personnel

150

Nurse or Other Personnel:

What she says	Verbal message	Exact content
How she says it	Nonverbal behavior	Facial expression
		Body position descriptors
		Scale of touch
		Verbal speech substitutes
		Patterns of movement as defined by spatial parameters of communication

Nurse Action:

What she does	Hygienic care
How she does it	Procedures, observation and measurement, treatment, etc.

Sociocultural Factors	Can be identified from chart information
	Can be identified through personnel records

Figure 4.   Guidelines for transcribing videotape interactions.

In attempting to gain more understanding of patterns of movement of patient and personnel, we have consulted with experts in dance. The principles of movement are presently being investigated for their applicability to this study. For example, the elements of movement are described by Hutchinson (1970) in terms of: the part of the body that moves; the use of space (direction and level); the timing of the movement (fast or slow); the dynamics, the texture of the movement (strong and light); and pattern or flow in the movement (bound or free).

Basing our work on these ideas, we have defined the spatial parameters for communication in terms of zones. The areas of space are defined in four zones, which allow the charting of movement of personnel.

The development of a methodology to examine the nurse-patient-physician communicative interaction process by describing nurse state, patient state, and physician state at a given point in time should provide insight into essential questions in nursing. We can still obtain answers to the questions of more classical studies such as numbers, durations, and frequency of interactions. More importantly, we can begin to examine the communicative interaction process by identifying the observed effects of the message, both verbal and nonverbal, upon the changing states of the participants as they move toward goal achievement. It is not possible to do this by using present methodologies which force the investigator to examine interaction events as single entities rather than a process.

One unresolved question which has stimulated much interest has been, the way patients define relationships with health care workers. In every communication the participants offer to each other definitions of their relationships or seek to determine the nature of the relationship. Similarly, each responds with his definition of the relationship which may confirm, reject, or modify that of the other. If the process does not stabilize, the relationship eventually is dissolved. Stabilization of relationship definition has been called the rule of relationship (Jackson, 1965). It is a statement of the redundancies observed at the relationship level, even over a diverse range of content areas. The rule may regard symmetry or complementarity or some aspect of relationship.

In psychotherapy, guidance and counseling, and social casework the rules of the relationship are usually defined prior to any exploration of the client problems or feelings. This is not generally true in a hospital setting, and questions arise about just what the rules of relationship in a hospital are. What is the effect on the patient if he is unable to define his relationship with either the nurse or physician? Or, what is the effect on the patient if he must continually attempt to define a relationship because numerous nurses and other workers are responsible for his care? One of the patients in our study, a 30-year-old schoolteacher who was scheduled for open-heart surgery, interacted with 33 different health care workers during the two-hour periods of observations over a ten-day time period. One cannot help but speculate about the amount of energy this required on the part of the patient.

It is believed that some patterns of interaction of nurse-patient-physician systems will lead to transaction or goal achievement while others will not. If we can identify these varying patterns of interaction, then we can start effective, economical, and therapeutic interventions.

REFERENCES

Abdellah, F. G. Frontiers in nursing research. *Nursing Forum,* 1966, *5*(1), 28-38.

———. Overview of nursing research 1955-1968: Part I, *Nursing Research,* 1970, *19,* 6-17.

Allport, G. W. *Pattern and Growth in Personality.* New York: Holt, Rinehart and Winston, 1961.

Ashby, W. R. *An Introduction to Cybernetics.* New York: Wiley, 1956.

Bartter, F. G., and Delea, C. S. Map of blood and urinary changes related to circadian variations in adrenal cortical function in normal subjects. *Annals, New York Academy of Science,* 1962, *98,* 969-983.

Bertalanffy, L. von. General system theory. *General Systems: Yearbook of the Society of General Systems,* 1956, *1,* 1-10.

Birdwhistell, R. L. *Introduction to Kinesics: An Annotation System for Analysis of Body Motion and Gesture.* Louisville: University of Louisville Press, 1952.

Chapman, J. S. Effects of different nursing approaches on phychological and physiological responses. *Nursing Research Report,* 1970, *5*(1), 1-7.

Cherry, C. *On Human Communication; A Review, A Survey, and A Criticism.* (2d ed.) Cambridge, Mass.: M.I.T. Press, 1966.

Dance, F. E. X. *Human Communication Theory; Original Essays.* New York: Holt, Rinehart and Winston, 1967.

Davitz, J. R. A review of research concerned with facial and vocal expressions of emotion. In J. R. Davitz (ed.), *The Communication of Emotional Meaning.* New York: McGraw-Hill, 1964.

Diers, D., and Leonard, R. C. Interaction analysis in nursing research. *Nursing Research,* 1966, *15,* 225-228.

Dittmann, A. T. The relationship between body movements and moods in interviews. *Journal of Consulting Psychology,* 1962, *26,* 480.

Ekman, P. Body position, facial expression, and verbal behavior during interviews. *Journal of Abnormal Social Psychology.* 1964, *68,* 295-301.

———. Communication through nonverbal behavior: A source of information about an interpersonal relationship. In S. Tomkins, and C. Izard (eds.), *Affect, Cognition, and Personality.* New York: Springer Publishing Co., Inc., 1965.

Howland, D., and Colson, H. Introduction to Systems Research. Unpublished manuscript, Ohio State University, 1970.

Hutchinson, A. *Labanotation, the System of Analyzing and Recording Movement.* New York: Dance Notation Bureau, 1970.

Jackson, Don D. The study of the family. *Family Process,* 1965, *4,* 1-20.

Johnson, B. A., Johnson, J. E., and Dumas, R. G. Research in nursing practice — the problem of uncontrolled situational variables. *Nursing Research,* 1970, *19,* 337-342.

Johnson, B. S. Relationships between verbal patterns of nursing students and therapeutic effectiveness. *Nursing Research,* 1964, *13,* 339-342.

Kaufmann, M. A. Autonomic responses as related to nursing comfort measures. *Nursing Research,* 1964, *13,* 45-55.

King, I. M. A conceptual frame of reference for nursing. *Nursing Research,* 1968, *17,* 27-31.

Leventhal, H. and Sharp, E. Facial expressions as indicators of distress. In S. Tomkins, and C. Izard (eds.), *Affect, Cognition and Personality.* New York: Springer, 1965.

Meyers, M. E. The effect of types of communication on patients' reactions to stress. *Nursing Research,* 1964, *13,* 126-131.

Miller, J. G. Living systems: Basic concepts. *Behavioral Science,* 1965, *10,* 193-237.

Morris, C. W. Foundations of the theory of signs. In O. Neurath, R. Carnap, and C. W. Morris (eds.), *International Encyclopedia of Unified Science, 1*(1). Chicago: University of Chicago Press, 1938.

Pride, L. F. An adrenal stress index as a criterion measure for nursing. *Nursing Research,* 1968, *17,* 292-303.

Ruesch, J. General theory of communication in psychiatry. In S. Arieti (ed.), *American Handbook of Psychiatry,* vol. 1. New York: Basic Books, 1959.

————. Clinical science and communication theory. In F. W. Matson and A. Montague (eds.), *The Human Dialogue.* New York: Free Press, 1967.

Sainsbury, P. Gestural movement during psychiatric interview. *Psychosomatic Medicine,* 1955, *17,* 458-469.

Satir, V. *Conjoint Family Therapy: A Guide to Theory and Technique.* (Rev. ed.) Palo Alto, Cal.: Science and Behavior Books, 1968.

Shannon, C., and Weaver, W. *The Mathematical Theory of Communication.* Urbana, Ill.: University of Illinois Press, 1949.

Smith, D. M. Myth and method in nursing practice. *American Journal of Nursing,* 1964, *64 (2),* 68-72.

Watzlawick, P., Beavin, J., and Jackson, D. D. *Pragmatics of Human Communication.* New York: Norton, 1967.

Wiener, N. *Cybernetics.* (2d ed.) New York: Wiley, 1961.

Wiens, A. N., Thompson, S. M., Matarazzo, J. D., Matarazzo, R. G., and Saslow, G. Interview interaction behavior of supervisors, head nurses, and staff nurses. *Nursing Research,* 1965, *14,* 322-329.

CHAPTER 13

Nurse-Patient-Physician Interaction in Caring for Patients in Cerebral Stress

Ann J. Buckeridge

The increasing number of persons entering the hospital for the treatment of central nervous system diseases places increasing demands upon the hospital resources. These demands are particularly complex when the patient is seriously ill because of brain pathology. The refinement of radiologic techniques — cerebral angiography and the ultrasonic and radioactive brain scans, for example — have expanded the physician's physical resources for definitive diagnoses and therapy. The electronic temperature regulator and the positive-pressure respirator are technological innovations that assist the patient during a period of severe cerebral stress. Chemical agents such as the cortisone derivatives and the phenothiazine preparations are utilized as adjuncts in the control and/or prevention of the inflammatory response that accompanies cerebral pathology. All of these resources have contributed to the expansion of the patient-care resources for the neurological patient.

Some hospitals have extended their physical resources to include intensive care areas for the neurologic patient and the acute stroke patient. Other hospitals accommodate those patients in the general medical-surgical care areas. Still others use a combination of intensive care and general care facilities, assigning the patient to that area which is best equipped with both human and physical resources to meet the individual patient's medical and nursing care needs.

Patterns of care for the neurologic patient are continually being modified to include such bioelectronic devices as physiological systems monitors and life-support systems. The array of biochemical procedures intended to provide precise clinical information, such as blood gas levels and central venous pressure measurements, continues to expand.

In addition to the need for expertise in medical practice, the neurologic patient-care service requires nursing resources that are capable of responding to a wide range of pathophysiologic and psychosocial needs of patients who are in stress because of their cerebral instability. The nurse who cares for the seriously ill neurologic patient is expected to monitor this patient's vital systems and to provide appropriate nursing interventions, which are directed not only toward the specific disorder but also toward the maintenance of all body systems placed in jeopardy during the period of stress. As the nurse attempts to meet these expectations, she may also be responsible for the operation of one or more pieces of complex machinery such as the hypothermia machine and the positive-pressure respirator. The nurse and the patient may be surrounded by such clinical paraphernalia as intravenous infusion equipment and tracheal and gastrointestinal suctioning apparatus, and by a variety of human resources in the patient's environment.

In this complex situation, decisions about appropriate nursing acts will be influenced by all of these constraints, as well as by the nurse's observation and measurement skills and her knowledge of the patient's particular clinical condition. The nurse's ability to select the best nursing action from an array of many possible actions will greatly influence effectiveness of nursing care.

In an effort to make these decisions, the nurse relies upon a number of information sources. The standards of patient care dictated by the medical and nursing services, as well as professional practice standards, will influence the decision process. The directions of the physician for such specific monitoring and regulating activities as the measurement of the patient's vital signs, cranial nerve function, level of consciousness, and neuromuscular behavior are information sources that are useful to both the physician and the nurse in evaluating the treatment regimen. The observed responses of the patient to all hospital activity, especially that involving his particular care program, provide still another information source from which to draw nursing inferences.

The purpose of this paper is to describe, within a conceptual framework of an adaptive systems model, what patient-state variables the nurse responds to while caring for the patient in cerebral stress; how the nurse measures and arrives at a value for the variable measured; and the patterns of nurse-patient-physician interaction which can be identified for systematic study. The data to be examined relate to four patients who were part of a pilot study conducted at The Ohio State University.

At the onset of this study we proposed that these data would support the observations made by the investigator in the neurologic care and the intensive care areas. These observations led to a set of general questions regarding how the nurse functions in this area of patient care.

We observed that the nurse functions differently when caring for the patient in the intensive care area than she does when caring for those in the general care area. Nursing functions appear to expand and contract in relation to the availability of the physician in the patient's immediate environment. When the patient is in the intensive care area, the physician not only functions as the controller but also participates in the monitoring and regulating activities. In the intensive care area the physician provides a high degree of patient surveillance, during which time he not only observes the patient but appears to measure some patient-state variables, such as level of consciousness, neuromuscular behavior, and response to painful stimuli. The nurse in this area appears to perform the prescribed monitoring and regulating activities and communicates these values to the physician on request or through the clinical record.

In the general care area the nurse assumes responsibility for essentially all of the monitoring and regulating activities and appears to modify these in response to changes in the patient-state values, regardless of the prescription of the physician. In this patient-care setting, the nurse communicates information about specific patient behaviors to the physician in several ways. Some of the information is recorded on the patient's record and is read by the physician at some time distant to the time of measurement. Some is reported to the physician via the telephone immediately following the time of measurement. There appears to be a pattern in both the measurement and the reporting of patient behaviors which can be related to the way the particular behavior is measured. For variables that can be measured with the nominal or the ratio scale, that is, the presence or absence of a response to a particular stimulus, or the blood pressure, pulse, and respiration rates, the measurement and reporting pattern reflects the prescribed schedule of the physician. If the order reads "take vital signs every 15 minutes," these values will be measured and recorded as ordered. Behaviors which cannot be so measured, that is, the level of consciousness, verbal activity, and spontaneous muscular-skeletal movement, are given a value presumably arrived at by the nurse's subjective estimation. Descriptive terms such as "the patient is confused," or "disoriented," or "semicomatose," are used to convey information about the level of consciousness. "Incoherent", "rambling," and "nonresponsive" are terms used to describe verbal behavior. Nurses use such terms as "restless," "combative," and "purposeless movements" apparently to convey a value for neuromuscular patient-state behaviors. The nurse's recording and/or reporting of these behaviors and their estimated values has an irregular pattern.

Our observations of nurses in the general neurologic patient-care setting have identified differences in the performances of individual nurses, a difference that is related to an overlapping of the nurse's monitoring and regulating activities. The experienced nurse appears to combine these activities. Her behavior in the monitoring function can be of two types: (1)

activities intended to obtain information about the spontaneous behavior of the patient, and (2) activities intended to obtain specific values for one or more patient-state variables. Examples of the first type of behavior measured by the nurse are the patient's response to the activity in the environment; his ability to participate in the activity; and his ability to communicate, verbally or nonverbally, such disturbances in his internal functioning as urinary and bowel urgency, thirst, and physical discomfort. This type of monitoring activity is referred to in the clinical neurologic literature as the general neurologic examination (DeJong, 1967; Locke, 1966). When the nurse infers that there is a disturbance in this behavior, as reflected in a change in the previously observed behavior, she will perform what is referred to as the formal neurologic evaluation. The second type of information seeking by the experienced nurse involves a series of purposeful stimuli intended to obtain discrete information for a particular patient-state variable, or specific groups of variables. This purposeful monitoring behavior appears to be unrelated to the prescribed monitoring schedule.

Throughout this formal evaluation the experienced nurse will alter her monitoring activity in a way that can change the value of the variable being measured, combining verbal and tactile stimuli to elicit a different response from the patient. She will increase the intensity of one or both of these stimuli, that is, the decibel level of the verbal message and/or the noxious character of the tactile stimuli. She will bombard the patient's sensory input channels until she infers that she has obtained a value that is compatible with the desired state. This desired state is usually an inferred state on the part of the nurse. For example, the initial response to stimuli intended to measure the patient's awareness of his environment may be an incoherent mumble. Repeated stimuli, combining both verbal and tactile intensity, will elicit a coherent but inaccurate verbal response. At this point in the process the nurse may infer that this is the desired level of consciousness for this particular patient at this moment in time. She usually does not report any of the above activity unless the patient's level of consciousness is deteriorating. Similar tactics have been observed when the nurse apparently infers a difference between the observed response and the desired response of the patient to a disturbance in his respiratory rate and rhythm. Upon noting an irregular rate or character of the patient's respirations, the experienced nurse has been observed applying this combination of verbal and tactile stimuli until there is an observable change in the direction of regularity. Again, no report is made unless the combined stimuli are not successful in righting the system.

These observations in the real world of neurologic patient care prompted the following questions: What is the essential information a nurse needs to infer a disturbance in the patient's neurologic behavior, and how can this information be obtained and communicated to the other members of the patient care team? What are the regulatory nursing actions this nurse must have at her disposal to bring the patient to and maintain him in the desired state of cerebral stability? Is there an *observable* difference between the types and number of monitoring and regulating activities that can be related to the desired state as prescribed by the controlling function? (The intent here is to identify differences in nurse behavior that can maintain the patient within the desired limits of stability. That is, the experienced nurse not only monitors and regulates as prescribed by the physician but uses other nursing strategies that keep the patient in this desired state.)

To answer these and other questions, we went first to the nursing and medical literature in an effort to find a baseline of essential information needed by the nurse in this area of practice. Our search revealed a general agreement that the nurse's primary function is one of monitoring the patient's neurologic state, interpreting this state in terms of some measurements of physiologic stability, and intervening when the patient goes outside the medically prescribed limits (Carini and Owens, 1970; Fuerst and Wolff, 1969; Shafer, Sawyer, McCluskey, and Beck, 1967). These sources do not clearly define how the nurse adapts her performance to the changing state of the patient. In general, nursing texts focus on the more common neurologic diseases, the diagnostic and treatment procedures, and medical knowledge that is relevant to nursing. All of these publications elaborate, to a greater or lesser degree, on the modifications of the general nursing care principles and concepts relevant to the maintenance of all the body systems placed in jeopardy during the period of cerebral stress. Read (1970) emphasizes the nurse's need for a comprehensive understanding of the normal physiology of the central nervous system, as well as a knowledge of the relevant principles from the biophysical and behavioral sciences.

The necessity for the nurse who cares for the seriously ill neurologic patient to be familiar with the technologic innovations used in the clinical management of this patient is well documented in recent publications (Edelstein, 1966; Gardner, M., 1968; Hickey, 1965; Martin, 1965; Price, 1967; Tarrent, 1966). Several clinical neurology texts that are directed toward the medical reader cite the significance of the observation and measurement function of the patient-care system and the interrelationships between several critical patient-state variables (DeJong, 1967; Gardner, E. D., 1968; Locke, 1966; Plum and Posner, 1966; Stern, 1966).

We found no studies of the information-decision process or the nurse's adapting to the patient's demands, although there is considerable interest in the literature for the development of research methodology to find out how nurses make decisions as they provide nursing care (Abdellah and Levine, 1965; Hammond and Kelly, 1966; McKay, 1969; Schlotfeldt, Quint, Conant, and Cleland, 1967; Wald and Leonard, 1964). During the past several years, a series of articles describing the research efforts of the Adaptive Systems Research Group (ASRG) of The Ohio State University has been published in the *American Journal of Nursing* and *Nursing Research*, as well as in the literature of other professional disciplines (Howland, 1963a, 1963b, 1970; Howland and McDowell, 1964; Howland and Wiener, 1963; Pierce, 1969; Silver, 1965).

An adaptive systems model seems promising in the development of new insight into how nurses make decisions while caring for the seriously ill neurologic patient. My experience as a practitioner and as an educator in this area of medical-surgical nursing does not support the usefulness of the disease model, that is, knowledge of the pathophysiology of the nervous system. The scope of this clinical knowledge is complex and it exceeds that needed by the practicing nurse to provide patient care. Medical and biophysical literature has not identified a baseline of information needed by all levels of nursing personnel. Our survey of the nursing texts in an attempt to obtain this baseline information revealed that the primary function of the nurse is to monitor the patient's neurologic behavior and to intervene with appropriate nursing actions when this behavior is outside the prescribed limits set by the physician. However, when we observe nurses who are identified by their physician and nursing colleagues as expert practitioners, the patterns of nursing activities do not conform to those prescribed in the literature.

The organization of neurologic nursing content within the theoretical framework of Selye's (1956) stress model and the concepts of stress of other biological and behavioral scientists (Guyton, 1966; Vickers, 1959) have led to the growing awareness of the need for a similar conceptual framework for nursing education. We believe that the conceptual framework of The Ohio State University ASRG can describe the real world of neurologic patient care and provide empirical data for patterns of nursing behavior which can be examined for significant areas of content for all levels of nursing education.

The goal of this inquiry is to gain an understanding of how the nurse responds to changes in the patient's behavior and how the patient responds to the behavior of all human resources in his environment, specifically the nursing resources. We believe that by observing, recording, and storing continuous data on the nurse-patient-physician interaction as it transpires in the care of the patient we can analyze these data to detect patterns in the nursing behaviors which will support the following hypotheses:

1. Nurses measure and record the patient's vital signs in accordance with the prescription of the physician, regardless of the change in the value(s) of these variables.

2. There is a difference in the amount and kind of information the nurse seeks when these variables begin to go outside the limits set by the physician.

3. The patterns of observation and measurement activities of the nurse in the intensive care area will differ from those she uses in the general care area.

4. Nurses respond differently to patient-state variables which cannot be measured quantitatively, such as level of consciousness and verbal behavior, than they do to those which can be so measured.

5. The amount and kind of stimuli in the patient's environment, both human and physical, will have a measurable relationship to the patient's ability to respond appropriately to those stimuli. The more purposeful the stimuli, the better able the patient is to stay within the desired state.

We believe that the experienced nurse in the neurologic patient-care setting responds to those neurologic behaviors of the patient which lead her to infer that a change has occurred in the value of one or more patient-state variables in the direction of disequilibrium. She intervenes with a nursing act(s) whenever this inference indicates a disturbance in the patient state which can be reduced by the regulatory action. The nurse obtains the patient-state information through observation and measurement activities. The more information the nurse is able to obtain, the better she can maintain the patient in the desired state (Howland and McDowell, 1964).

The Ohio State University ASRG is an interdisciplinary team concerned with the measurement of patient care in the health care system (Howland, 1963a, 1963b, 1970; Howland and McDowell, 1964; Howland and Wiener, 1963; Pierce, 1969; Silver, 1965). This research is based on the rationale that the nurse-patient-physician triad is the basic subsystem of the hospital patient-care system. The triad is viewed as an adaptive system that has as its purpose the maintenance of homeostasis for the patient at a time when he seems unable to perform this vital function unaided. A system is defined as a collection of man-machine components connected by a communication network which has purposeful goal-directed behavior. An adaptive system is one which takes action to restore itself to a specific state. There are a number of homeostatic mechanisms in nature. Science and technology have duplicated the temperature control regulatory mechanism of the hypothalamus in the mechanical thermostat universally used in the heating and air conditioning industry. The writings of Claude Bernard in the last century (1865; translation, 1957) and Cannon (1932) and Selye (1956) in this century discuss the adaptive capabilities of the human organism.

The triad model as conceptualized by the ASRG is descriptive of the interaction of the components of the nurse-patient-physician patient-care system. The present model (Figure 1) consists of a series of functional components. The controller specifies the limits within which variables are to be maintained for the patient Ys. Limits are based on information about the individual patient. They are subject to change as the patient's condition varies. The patient is a signal source generating such patient-state information as vital signs, response to stimuli, and level of consciousness. The comparator compares the actual value of the variable measured by the monitor with the desired value specified by the controller, and notes the difference that may exist. The regulator selects an action intended to reduce this difference in the direction of the limits set by the physician. The resources used by the regulator are represented by the Xs in the model. The patient responds to these and generates information which may or may not be detected by the monitor. The information that is detected becomes the data from

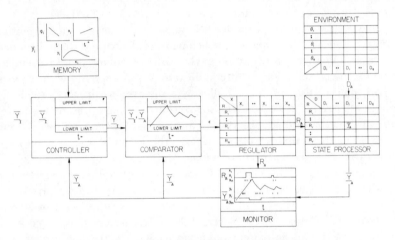

Figure 1.   Cybernetic models of adaptive systems.
(Reprinted from D. Howland, "Toward a Community Health System Model," in *Systems and Medical Care,* ed. A. Sheldon, S. Baker, and C. P. McLaughlin. Copyright M.I.T. Press, 1970)

which decisions are made. A number of alternate actions may result from a decision. The decision may be to take no action. Two types of information are needed in this information-decision process: first, information about the specific patient performance prescribed by the physician and the actual performance measured by the monitor at an instant in time; and second, information about strategies, or alternate choices of action, available to the nurse to reduce the difference (Howland and McDowell, 1964).

An illustration of the system behavior is activity in the intensive care unit immediately following the patient's brain surgery. The patient's blood pressure is an essential variable in the reestablishing and maintaining of his cerebral equilibrium. The physician, functioning as the controller, prescribes the limits of the systolic and diastolic range. Values of this variable are measured and recorded at time intervals stipulated by the physician. The values, monitored by the nurse, are compared with the prescribed limits, and, if there is a difference, the medical team undertakes a regulatory action. Certain regulatory actions are within the province of the nurse, others are the responsibility of the physician. For example, the nurse may note that the values are traversing toward the outer boundaries of the limits and may then reduce the noise level, change the patient's position, increase the frequency of the monitoring activity, or notify the physician. The doctor, in turn, may prescribe a medication or select some other regulatory tactic to bring the blood pressure back to the desired state.

Certain physiologic measures, such as body temperature, systolic and diastolic blood pressure, and the rate and character of the pulse and respirations, are indicators of the patient state for all hospitalized individuals. Certain other measurements are directly related to the pathology, such as the ratio of fluid intake and urinary output for the patient with kidney failure. For the patient with cerebral pathology, the nurse and physician measure the above vital signs and, in addition, measure the patient's response to verbal and tactile stimuli, his motor-sensory coordination, the pupillary symmetry and response to light stimuli, and his level of consciousness (DeJong, 1967; Evans, 1963; Plus and Posner, 1966). The initial phase of our study explored this set of patient-care activities and used the triad model as the means of describing what is happening between the patient, nurse, and physician when the systems goal is to reestablish and maintain the patient in cerebral equilibrium.

In the Adaptive Systems Research Group we conducted a pilot study to test the feasibility of the triad theoretical framework and the methodology for the description of the nurse-patient-physician interaction in the area of neurologic care. The long-range objective of the research is to improve patient care. The immediate objectives are to identify significant patient-state variables, to identify the nurses' responses to these variables, and to identify patterns of nurses' responses to the patient-state variables which maintain the patient in the desired condition.

The research is somewhat circular in concept, beginning and ultimately returning to the clinical areas where neurologic patients are receiving care. We began by collecting time-series data on patients who were seriously ill because of brain pathology. We anticipated that the study will expand to include the triad interaction in the recovery and rehabilitation phases of central nervous system illness (Figure 2). We elected to begin at the critical phase of illness because we believe that in this period of physiologic disequilibrium the patient-state variables and the resource variables are the most complex and the most difficult to measure.

We are now at the second phase of the research in which we hope to draw from the pilot study testable hypotheses that can lead to the third phase, an exploratory study to identify essential variables through the testing of these hypotheses. Phases four and five are essentially a continuation of research into a quasi-experimental design in which we will use the data to simulate the real world for training purposes. Assuming that a body of nursing knowledge will evolve through a thorough description of the real world of patient care, we expect that this research will identify a body of nursing content that will become the domain of the academically well-prepared nurse practitioner. We also hope to find a component of knowledge that is essential for the nonprofessional worker in this area of nursing. If we can systematically organize all this knowledge into educational programs for both levels of workers, our long-range goal — the improvement of patient care — should be realized.

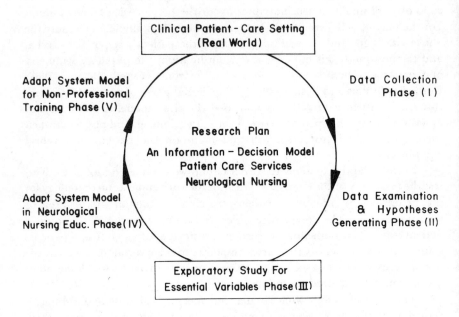

Figure 2.   Nurse-patient-physician interaction in the care of the patient in neurologic stress.

A timetable for the initial phase of the research included an orientation to the ongoing activities of the ASRG staff and an opportunity to use the data-collecting instrument in the patient-care setting. Since descriptive studies start with time-series data that describe the system, we went to the neurologic patient-care areas for our patient population. We observed continuously for the first 48 hours four patients who were experiencing cerebral disequilibrium following brain surgery. One patient remained in the surgical intensive care unit for the entire observation period. Three patients spent a portion of the postoperative period in intensive care and were transferred to the neurologic general-care area during the observation periods. Parenthetically, we are interested in what patient-state values are used to decide that the neurologic patient is ready to be moved from the intensive care area to the general care area.

Trained observers of the ASRG staff who were nurses themselves recorded in narrative form all the observable interaction in the patient's environment. Values were recorded for specific measurements such as vital signs, pupil response, and body temperature, as well as for all the patient's observable motor activity. The observers also recorded hospital resource activities, such as measuring of intravenous infusion rate, measuring urinary output, and administering specific medications and treatments. Verbal activity of everyone in the patient's environment was recorded verbatim. At the time of the pilot study these data were categorized as verbal messages to and from the patient without further breakdown as to message content.

After the reduction, coding, and keypunching of these data, we decided to analyze in depth the data about one patient. The data described what happened to an elderly gentleman who came to the hospital in cerebral stress due to a subdural hematoma. Our observation period began immediately after the surgical procedure. The patient remained in the intensive care unit for eight hours and was then transferred to the general care unit, where he remained through the next 36 hours of the observation period. The data were then prepared for analysis.

The ASRG staff has developed a number of computer programs for data analysis. The initial program selected for our data provided frequency counts for all variables, and the computer was programmed to count each time a specific event occured (the times a measurement appeared, the visits to the patient's bedside by a member of the medical and nursing staff, and the frequencies of several other events). We selected specific events that appeared repeatedly in the data for further study. For example, we hypothesized that the patient-state variables measured by a discrete value scale, such as the vital signs, would appear at regular intervals in the data and that variables estimated by the observer, such as the verbal and tactile responses, would appear in an irregular pattern.

In an effort to examine specific relationships between patient-state and resource activity variables, we cross-tabulated the frequency of personnel activity according to type (Table 1). It is interesting to note that the professional nurse shows the highest incidence of patient-care activity in all categories, with the major portion of her activity in the monitor component of the system. The professional nurse also is involved in numerous care activities that require her to be in visual, auditory, and tactile communication with the patient. There was no apparent difference between the distribution of nursing activities in the intensive care area and that in the general care area. The major portion of the physician's activity took place in the intensive care unit.

Table 1. Frequency of Patient Care Acts: According to Type and Human Resources in Environment (48-Hour Observation)

Environment Resource	Observe and Measure	Medicine and Treatment	Comfort and Safety	Verbal Stimulus	Total 48 Hrs
Physician	23	1	1	10	35
Professional Nurse (RN)	149	96	40	47	332
Practical Nurse (LPN)	68	9	11	13	101
Nurse Aide	76	2	30	8	116
Visitors	0	0	1	15	16

Table 1 also points up an interesting distribution in the area of verbal stimuli to the patient. This may, however, reflect a limitation of the data reduction procedures, during which all verbal interaction was classified as verbal messages either to or from the patient. Messages that the observer interpreted as being directly related to the measurement of the patient's level of consciousness were coded as verbal stimuli. Further examination of narrative data revealed that there was considerable verbal activity in the patient's immediate environment in both the intensive care area and the general care area. The data raise a question about the relationship between the total activity, including the verbal activity, and the ability of the system in maintaining the patient in the desired state.

Studying further this verbal interaction in the patient's environment, we reexamined the narrative data and cross-tabulated the frequency of the verbal messages to the patient according to message content and message sender (Table 2). We are now examining these data to identify relationships between the type of verbal message and the change, if any, in the patient state after the particular verbal stimulus.

We are particularly interested in the relationships between the verbal and touch activities of the nurse and the patient's ability to respond to these in an appropriate manner. Examples of auditory stimuli the nurse might give are questions related to the patient's awareness of his environment and instructions requiring a cognitive response. Examples of tactile stimuli are bathing, comforting, and nonpainful touch.

Our next computer program provided a visual display of combinations of patient-state and regulatory-action variables. Twenty-seven variables were included in a multivariable plot (Figure 3). The abscissa is calculated in real time, with this particular segment representing two hours of the total observation period. Each patient-state variable and its value is printed at the time of its measurement along the ordinate of the plot. The plus symbol indicates the presence of a response to the stimulus, and the zero sign indicates the absence of a response. Each hospital activity is identified by a subscript symbol representing the category of the person performing the activity. For example, when the professional nurse measures the patient's response to tactile stimuli, the subscript N is recorded on the printout. If both a nurse and a physician perform the same activity within a ten-minute interval, both an N and a P will appear in that time segment.

These particular time-sequence data display a change in three patient-state variables and a resulting change in the resource activities. At the time this activity was taking place the physician's prescription required the observation and measurement of the patient's vital signs and response to tactile and light stimuli every thirty minutes. This figure only displays the value of the systolic blood pressure from all the values recorded for the total complex of the vital signs. There is a change in the direction of the value of the systolic

Table 2. Frequency of Verbal Messages According to Message Content and Sender (48-Hour Observation)

Message Sender	Motor Command	ORIENT Time Place Person	INFORM Care Activity	INFORM Behavior Activity	INFORM Environment Activity	Total Messages
Physician	14	10	1	10	0	35
Prof. Nurse (RN)	75	101	46	43	3	268
Pract. Nurse (LPN)	14	3	2	0	2	21
Nurse Aide	8	3	10	11	4	36
Visitors	0	10	0	20	5	35
Total	111	127	59	84	14	395

Patient Behaviors (Y's)
Systolic BLP

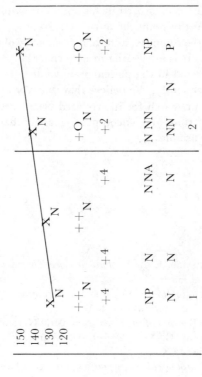

150					
140		X_N			X_N
130	X_N		X_N		
120					
Pupil Response (Light)	++N	++N		+O N	+O N
Moves Extremities	+4	+4	+4	+2	+2
Hospital Activity (X)			NNA		
Verbal Stimulus	NP	N		N NN	NP
Tactile Stimulus	N	N		NN	P
Time (Hrs)	1		2		3

Figure 3. Relationship of X and Y values for one patient: Sample of two of forty-eight hours.

blood pressure with each measurement recorded. At the beginning of the second hour, two additional changes in value appear—a change in the pupillary response to the light stimuli and a change in the patient's response to the tactile stimuli as measured by his apparent inability to move two of his four extremities. At this point the nurse increases the frequency of measuring the verbal and tactile response, and there is an entry of the physician's symbol in the activity row. It is interesting to note that the additional measurement activity is not reflected in the patient-state values, nor is the blood pressure measured more frequently. We believe that this type of multivariable display can be useful in our search for interrelated events occurring in a relatively rapid sequence in time and encompassing a number of patient-state and hospital-resource variables.

REFERENCES

Abdellah, F., and Levine, E. *Better Patient Care Through Nursing Research.* New York: Macmillan, 1965.

Bernard, C. *An Introduction to the Study of Exceptional Medicine.* Translated by Henry C. Greene. New York: Dover, 1957.

Cannon, W. B. *The Wisdom of the Body.* New York: Norton, 1932.

Carini, E., and Owens, G. *Neurological and Neurosurgical Nursing.* (5th ed.) St. Louis: C. V. Mosby, 1970.

DeJong, R. *The Neurological Examination.* (3rd ed.) New York: Hoeber Medical Div., Harper & Row, 1967.

Edelstein, R. R. Automation: its effect on the nurse. *American Journal of Nursing,* 1966, *66*, 2194-2198.

Evans, J. P. *Acute Head Injuries.* (2d ed.) Springfield, Ill.: Charles C Thomas, 1963.

Fuerst, E. V., and Wolff, L. *Fundamentals of Nursing; the Humanities and the Sciences in Nursing.* (4th ed.) Philadelphia: Lippincott, 1969.

Gardner, E. D. *Fundamentals of Neurology.* (5th ed.) Philadelphia: W. B. Saunders, 1968.

Gardner, M. Responsiveness as a measure of consciousness. *American Journal of Nursing,* 1968, *68*, 1034-1038.

Guyton, A. C. *Textbook of Medical Physiology.* (3rd ed.) Philadelphia: W. B. Saunders, 1966.

Hammond, K., and Kelly, K. Clinical inference in nursing. *Nursing Research,* 1966, *15*, 23-38.

Hickey, M. C. Hypothermia. *American Journal of Nursing,* 1965, *65* (1), 116-122.

Howland, D. A hospital system model. *Nursing Research,* 1963, *12*, 232-236. (a)

Howland, D. Cybernetics and general systems theory. *General Systems; Yearbook of the Society of General Systems,* 1963, *8*, 227-232. (b)

Howland, D. Toward a community health system model. Chapter 11 in A. Sheldon, F. Baker, and C. P. McLaughlin (eds.), *Systems and Medical Care.* Cambridge, Mass.: MIT Press, 1970.

Howland, D., and McDowell, W. E. The measurement of patient care: A conceptual framework. *Nursing Research,* 1964, *13,* 4-7.

Howland, D., and Wiener, E. L. The system monitor. In D. N. Buckner and J. J. McGrath (eds.), *Vigilance: A Symposium.* New York; McGraw-Hill, 1963.

Locke, S. *Neurology.* Boston: Little, Brown, 1966.

Martin, A. Care of a patient with cerebral aneurysm. *Americal Journal of Nursing,* 1965, *65,* 90-95.

McKay, R. Theories, models and systems for nursing. *Nursing Research,* 1969, *18,* 393-399.

Pierce, L. M. A patient-care model. *American Journal of Nursing,* 1969, *69,* 1700-1704.

Plum, F., and Posner, J. B. *The Diagnosis of Stupor and Coma.* Philadelphia: F. A. Davis, 1966.

Price, D. Data processing—present and potential. *American Journal of Nursing,* 1967, *67,* 2558-2564.

Read, E. H. Nursing the patient having a problem resulting from disorders in regulation: Neural regulation. In I. L. Beland (ed.), *Clinical Nursing.* (2d ed.) New York: Macmillan, 1970.

Schlotfeldt, R., Quint, J. C., Conant, L. H., and Cleland, V. S. Research—how will nursing define it? *Nursing Research,* 1967, *16,* 108-121.

Selye, H. *The Stress of Life.* New York: McGraw-Hill, 1956.

Shafer, K. N., Sawyer, J. R., McCluskey, A. M., and Beck, E. L. *Medical-Surgical Nursing.* (4th ed.) St. Louis: C. V. Mosby, 1967.

Silver, A. An investigation of information factors upon the performance of the human regulator. Unpublished doctoral dissertation, Ohio State University, 1965.

Stern, W. E. Cranialcerebral injuries. In A. M. Nahum (ed.), *Early Management of Acute Trauma.* St. Louis: C. V. Mosby, 1966.

Tarrent, B. J. Automation: Its effect on the patient. *American Journal of Nursing,* 1966, *66,* 2190-2194.

Vickers, G. The concept of stress in relation to the disorganization of human behavior. In J. M Tanner (ed.), *Stress and Psychiatric Disorder.* Oxford: Blackwell Scientific Publications, 1959.

Wald, F. S., and Leonard, R. C. Towards the development of nursing practice theory. *Nursing Research,* 1964, *13,* 309-313.

CHAPTER 14

Bridging the Gap between Systems Theory and Research

Grayce M. Sills

Paul Bohannon in a recent article titled "Beyond Civilization" (1971) writes that an intense crisis threatens the world in which we live. Bohannon is not alone; across the nation, within and without academia, the same cry is sounded. So the Oggian leprechauns in our midst search for the pot of gold that will restore or maintain a former or some present state. At a more serious level, it may be true that civilization as we know it is likely near disaster — disaster that may spell the end for this civilization, like others before it. Or, as an alternative hypothesis, we may be near to one of those giant evolutionary steps that we can only dimly comprehend at this moment in time. Bertrand Russell (1955) has said:

Science, whatever unpleasant consequences it may have by the way, is in its very nature a liberator, a liberator of bondage to physical nature and, in time to come, a liberator from the weight of destructive passions. We are on the threshold of utter disaster or of unprecedentedly glorious achievement. No previous age has been fraught with problems so momentous; and it is to science that we must look for a happy issue (p. 17).

What may be clear now, however, is that there is a recognition of and concern for the near disastrous state of affairs in the solution of the myriad human problems that beset us on all sides. In thinking about this great concern, I have been impressed by the tremendous increase of interest in the past ten years in general systems theory, systems concepts, and a systems perspective. In one sense this resembles the "convergence phenomena" in natural disaster noted so often about behavior under disaster. Still, while such convergence behavior in times of disaster does create problems of traffic control, crowd control, and the like, it also provides a manpower pool which can function as resources in solutions of the problem. I do not wish to push the analogy too far, but it does seem to offer one possible explanation for the high level of interest in systems.

Several factors may have contributed to the process of the crisis as it has formed and even continues to develop. While all the factors are not known, neither can all those known be examined within the body of this paper. Hence, those with a more immediate bearing on the topic — the gap — will be dealt with.

The Ethical Issue

First, there is perhaps the near paralysis of the basic sciences — the hard sciences — to treat of their ethical issues, particularly those which deal with consequences. Certainly, since the Manhattan Project the sensitivity seems present, but the faint unspoken hope seems to be that someone (the social sciences?) will bail them out. Let us take an example: to be able to generate life in a test tube, or to be able to control the weather, are capabilities within either the present or near future. The assessment of total systemic consequences for humanity of either of these two outgrowths of hard scientific research has only begun to be faced. Thus, the right hand of science generates the "problems" for the left hand. The analogy to "iatrogenesis" in medicine is striking.

The Traditional Boundaries

Second, and related to the first, is the issue of traditional boundaries for knowledge. The "knowledge pie" is carved with an "artifactual knife" into the basic sciences (the pure) and the applied sciences (the impure?). The boundaries are kept intact by traditional patterns of university organization and the even more powerful sanctions of membership in the elite societies and associations of the scientists.

From time to time men have emerged who could seemingly transcend the boundaries of traditional disciplines, including Dubos, Grinker, Boulding, Miller, Lewin, and von Bertalanffy. Such men provide a vision and direction which offers hope to the beleaguered man, the maverick, the marginal man, those who walk between boundary lines, those who work in interface areas. It is these latter men and women who suffer in the day to day struggle of working to make the sciences responsive to real world problems. It is they who must sometimes feel disoriented and discouraged.

And it is so; the existence of one who works at and in the interfaces is a precarious one. The state of the marginal man has always been lonely, relatively unrewarding, and frustrating. Yet such a marginal position is often compensated for by the wide-angle view of reality which is thereby obtained. Thus, most systems people find themselves in this double-edged position. The dilemma has no easy out, nor does there need to be a pathway out. For it is in the midst of the marginality that one can obtain that view of which Kurt Lewin (1948) wrote, "the research worker can achieve this [bridging the gap from theory to reality] only if, as a result of constant intense tension, he can keep both theory and reality within his field of vision" (p. xvi).

One set of authors has commented:

We have developed a substantial body of theory and certainly a rich body of practice, but somehow our failure has been to provide the transformations and the bridging between the two. . . . We seem quite often to become lost at the crossroads of a false dichotomy; the purity and virginity of theory on the one hand and the anti-intellectualism of some knowledge-for-what adherents on the other (Bennis, Benne, and Chin, 1969, p. 4).

Fear of Science

Closely related to this is a third factor which merits attention. Brodey formulated the issue in a paper presented to his colleagues at the American Psychiatric Association convention in 1966 and spoke of the "fear of science." Speaking from a vantage point of marginality, he talked about the integration of man and machine and the reaching out of experts on both sides of the bridge:

This bridging is slowed by ignorance and by our own deficits as psychiatrists in finding ways of modeling our observations that would be more effective in communication. Our deficit, I believe, is due to the fear of science in which we psychiatrists lived — and I think justifiably so — in precomputer days, when there was no possibility of real time computation . . . that it is now time for us clinicians to come out of hiding and to leave behind the fear that science means to destroy our discipline in its complexity in order to make it measurable (Brodey, 1969, pp. 229-230).

Brodey's point is well taken and can be extended to include all who view themselves as practitioners in the sense of application of knowledge in the solution of human problems. Typically, practitioners hold values thought to be disparate from hard scientists. Hence, a reluctance to share and collaborate pervades the practitioner-scientist relationship. These qualities promote fear and mistrust and are inimical to the resolution of human concerns. Only as we are able to value and reward our marginal men from both sides will it be possible to foster the mutual development of theory that is even more useful to guide research to enhance practice.

I have implied a dichotomous theme. While it is true that to some extent the issues are dichotomous, it is truer that in the present state of transformation of theory to research the designs often are clouded by a lack of clarity about the theory itself.

Operational Style — A Way of Thinking and General Systems Theory

When one takes a closer look at the gap between systems theory and research, the development of an offspring of general systems theory becomes an issue of concern. The offspring has adopted general systems theory as a parent, but it is yet to be known whether or not they will become a family. Auerswald (1969) has been particularly apt in describing how this loose alliance has come about:

A relatively small but growing group of behavioral scientists . . . have taken the seemingly radical position that the knowledge of the traditional disciplines as they now exist is relatively useless in the effort to find answers for these particular problems. Most of this group advocate a realignment of current knowledge and reexamination of human behavior within a unifying holistic model, that of ecological phenomenology. The implications of this departure are great. Once the model of ecology becomes the latticework upon which such a realignment of knowledge is hung, it is no longer possible to limit oneself to the behavioral sciences alone. The physical sciences, the biological sciences, in fact, all of science, must be included. Since the people who have been most concerned with constructing a model for a unified science and with the ingredients of the human ecological field have been the general systems theorists, the approach used by behavioral scientists who follow this trend is rapidly acquiring the label of the "systems approach," although a more appropriate label might be the "ecological systems approach."

These terms are currently being used metaphorically to describe a way of thinking and an operational style. They do not describe a well-formed theoretical framework as does the term *general systems theory* (pp. 373-374).

Buckley (1968) says that the general systems theory perspective is promising for "organizing these valuable insights of the past and present," because the perspective embraces "a holistic conception of complex adaptive systems viewed centrally in terms of information and communication process and the significance of the way these are structured for self-regulation and self-direction" (p. 508).

Such is the state of the art. Until synthesis can take place, and it will, many will continue to fragment reality into artificial slices, but the concern for wholeness and unity continues to press from the real world. Hence, researchers must carefully reconsider the theory that informs the search. The task is enormous and not well fitted to a society that places high value on quick results and ready answers. In fact, a recent report suggests the urgency of the problem (Also RANN, 1971).

Systems and Subsystem Relationships

The most difficult tasks for the theorists in general systems have been to specify and name the interface linkages between systems and subsystems. This problem is confounded by the nature of the English language. Jay Haley has referred to this as the age of the hyphen, meaning that interactional relationships have had to be characterized in terms of two individual terms linked by a hyphen; for instance, input-output, dominant-submissive, and so on. What are only beginning to be developed are process terms, transactional terms which can be defined and then operationalized by researchers.

Brodey (1969) illustrates this conceptual necessity by inventing the terms "time gaining," "time driving," and "information time." Significantly, Brodey needs two words to conceptualize each process. He writes:

The bottleneck in making computer resources available to man is the need to interface their communication so that it will grow in the style of a real conversation — as exploratory conversation, as learning conversation, where the associations and experience of each will be available to the other. The innovative use of computers depends upon solving this dialogue problem. Man must adapt to the advantages bestowed on him by his inventions — fire, wheel, atomic energy, and computers, among many — or die.

As we learn to build computer programs that have a little more ability than those we know so far to enter into a rich and complex dialogue with man, we begin to learn a little better how to understand what we have hidden in our art and our practice — the science of information exchange (pp. 242-243).

Thus, while the conceptual language to deal with interacting processes is in its infancy of development, theorists are beginning to inform researchers of the crucial linkages. Such information must run both ways. Unless it does, the promise of general systems will become an empty pot at the end of the rainbow.

Further, such conceptual linkage is necessary to enable decision-makers to consider the system consequences for any alteration or change in the system state. For example, we may assume the level of analysis is a system and the hospital is the analytic unit at that level, and that many subsystems can be identified. One of these is the pharmacy subsystem, a component of the whole. This component had developed a "nearly perfect" operational method for drug administration. The rationale for the adoption of the new method was considered in terms of the interface with the system component of nursing. It was determined that adoption would enhance the functioning of the nursing component. Hence, the new operation of the subsystem was implemented. Previous failure to consider the effect on other intra-system components as well as failure to consider extra-system components characterized this decision. No one considered cost factors, interactional consequences with other systems, and such. Similar illustrations are probably widely available. No other existing theory except general systems can offer a framework within which such complex interfaces can be researched. A possible exception to this statement may be Gunnar Myrdal's (1968) theory of circular causation.

If researchers and practitioners can communicate to theorists about the gaps in the theory, then theoretical models can be translated into tactical strategies which can yield explanation, which in turn can lead to the type of prediction and control we so desperately need if we are even to develop into what Paul Bohannon (1971) calls Civilization II.

REFERENCES

Also RANN for behavior. *Behavior Today,* 1971, *2*(8), 2.

Auerswald, E. H. Interdisciplinary versus ecological approach. In W. Gray, F. J. Duhl, and N. D. Rizzo (eds.), *General Systems Theory and Psychiatry.* Boston: Little, Brown, 1969.

Bennis, W., Benne, K., and Chin, R. Introduction. In W. Bennis, K. Benne, and R. Chin (eds.), *The Planning of Change.* (2d ed.) New York: Holt, Rinehart and Winston, 1969.

Bohannon, P. Beyond civilization. *Natural History,* 1971, *80*(2), 50–67.

Brodey, W. M. Information exchange in the time domain. In W. Gray, F. J. Duhl, and N. D. Rizzo (eds.), *General Systems Theory and Psychiatry.* Boston: Little, Brown, 1969.

Buckley, W. Society as a complex adaptive system. In W. Buckley (ed.), *Modern Systems Research for the Behavioral Scientist.* Chicago: Aldine, 1968.

Lewin, K. *Resolving Social Conflict.* New York: Harper, 1948.

Myrdal, G. *Asian drama: An Inquiry into the Poverty of Nations.* Vol. 3. New York: Twentieth Century Fund, 1968.

Russell, B. Science and Human Life. In J. R. Newman (ed.), *What Is Science?* New York: Simon & Schuster, 1955.

PART FIVE

Systems Approach Applied to the Realm of Client and Practitioner

Earlier papers in this book dealt with the broad health care system and with one model that was used to study the system. The papers in this section examine the systems approach from the standpoint of both the client and the provider of health services in eight areas of interest. Each of the eight papers focuses clearly on the world of work in the health fields.

Meyer and Kolins discuss the studying of community health from the view of health provider and consumer and emphasize the interaction between these community subsystems. Considering psychosocial factors as inputs, they illustrate four major components of the health care process: recognition and acceptance of need; seeking care, and accepting or rejecting the "contract"; patient care process; and evaluation of outcome of need and of the quality of health care relationships. They suggest research on certain aspects of these components.

Passos discusses her views about model building in a practice field, especially her concern about bringing into focus the whole person, the client, whom the health professionals serve. She questions whether general systems theory is applicable to

179

psychological phenomena. Definition of system, she argues, is not always clear for either the client or the health care provider, and she believes that inclusion of goals in the system definition would be helpful. Passos describes a survey conducted to evaluate the quality of the output of the nursing service of an urban general hospital and to measure selected conditions under which care was provided, defining some of these conditions in terms of the client system and its environment.

Hess discusses how the systems approach can be used in planning continuing education for health professionals. By diagrams and examples he shows that the educational intervention can be scientifically derived from data related to quality of care and educational needs. The systems approach model makes provision for testing hypotheses related to particular health care problems within a health care system as defined by a hospital; thus, data can be collected to guide the planned educational intervention. Although his example pertains to medical education, the model can serve equally well for any of the health professionals.

Abbey shows how the systems approach is employed in clinical nursing studies of physiological controls. She discusses some aspects of general systems theory that are central to designing the research, but she also makes the point that the systems approach must be tied to the overall conceptual framework of the problem under investigation. The two projects for which pilot studies are in progress pertain to nursing management of temperature regulation. Her studies pertain to both the world of work and the world of the client; studies of this type are essential for articulation between these two worlds of the health services.

Smoyak's main argument is that theory is powerful, and she demonstrates how different findings are produced by using differing theories in looking at a social system. She indicates that so-called systems research in hospitals to date has not captured the data on interaction. She attributes the paucity of systems research in hospitals to the fact that systems theory and methods have not been sufficiently developed or translated for use in hospitals or the health fields. She describes a hospital study in which systems thinking can be applied, for the methodology does include analysis of nursing units in addition to other more traditional methods. In this situation sociologists studied the process of technological change and its implications when the hospital's administration introduced a computerized hospital information system. Differing philosophies of admini-strators, medical staff, and other hospital personnel were important in regard to implementing a change such as the hospital-wide computerized information system. Difficulties can also arise when major changes are introduced as the result of a unilateral decision. She reports some of the stresses resulting from the innovation as they were experienced by individuals and units within the system. In so doing, she discusses the adequacy of the socialization model for systems analysis, showing that with the application of a systems model the innovation would be considered a success, whereas the socialization model led to a conclusion that it was a failure.

Medical diagnosis is a crucial skill for practicing physicians and one about which medical educators are concerned as they prepare future doctors. The ongoing research by Schwartz and Simon on information processing and decision making in medical diagnosis should eventually lead to a better under-standing of the components of diagnostic skills, and therefore should be helpful to educators. The authors review succinctly the extensive literature bearing on their work in medical diagnosis, showing how they have built on earlier research. They outline their own explicit model of information processing procedures, which constitute medical diagnosis, indicating that they conceive of medical knowledge varying along two dimen-sions—breadth and structure. Their model divides the diag-nostic process into ten states with six subprocesses or operations involved in traversing through these ten states. They discuss some initial investigation using the model; preliminary findings have shown that knowledge structures can be identified and classified. They are hopeful that continued research on the diagnostic processes and their interaction will be of consider-able import in understanding complex cognitive functioning.

Zagornik shares her views on the application of the systems approach to the teaching-learning process. She sees it as a learn-ing tactic of considerable importance in developing two major cognitive skills of analysis and synthesis. She views the nursing process as a system of related processes and believes that under-standing of this process reflects the development of a systems ap-proach to nursing. A five-phase plan shows the steps she follows in teaching graduate students to use the systems approach in their studies of nursing and the health delivery services. These phases pertain to the black-box concept, nursing process and black-box relationships, nursing process and control systems relationships, conceptualization and development of "own" models, and application of the systems approach in studying a particular nursing problem. She shares students' reaction to utilizing the systems approach in their studies in the health fields.

Jelinek has done extensive research in the patient-care area, where he employs the systems approach to measuring and modeling the patient-care operation. His discussion pertains to performance of hospital patient-care services that are not performed by physicians and are not the direct responsibility of physicians. He presents a model and describes the factors involved in the patient-care system and then moves on to the problem of measuring performance of the system, looking at both quantity and quality measures. Through his input-output model, predictions can be made, and such information is vital in the management of the patient-care system. Jelinek is aware of the problems related to the lack of satisfactory measures of quality of care, the need for variable standards associated with each performance, and the need to identify factors that either determine or influence standards.

As an example of one aspect of management control, he discusses utilization of nursing personnel and diagrams the relationship between workload and staff size. The approach is based on classification of patients into categories that reflect varying care needs; some of the earliest work done in the area of categorization of patients according to nursing care needs was conducted in Army hospitals in the 1950s (Claussen, 1955; Nursing Service Management Research Team, 1952). Jelinek mentions some of the important factors to be considered in the review of relationships between staff and work load.

References

Claussen, E. Categorization of patients according to nursing care needs. *Military Medicine*, 1955, *116*, 209–214.

Nursing Service Management Research Team. Nursing Service Organization and Utilization Analysis. Report prepared for the Office of the Surgeon General, Army Medical Service, Valley Forge Army Hospital, 1952.

CHAPTER 15

Systems Approach to Studying Community Health

Ruben Meyer and Maura Kolins

Before discussing the content and process of community health studies, one necessarily must consider who is doing the studying and how their activities are organized. A systematic analysis of community health problems will only emerge from a systematic organization of research personnel. According to Murray (1971), there are three roles that must be represented in a health research effort: the technologist, the manager or planner, and the systems analyst. The technologists should ideally include persons who have knowledge and experience in the scientific base of the study area (professionals) and others who have practical knowledge of community activities and needs (consumers). Managers and planners include institutional personnel who are responsible for the system in which the technologist operates. Systems analysts are required to organize the inputs of the other two actors and synthesize a systematic approach and conclusion to the subject under study. These roles may be performed by one or more individuals, but it is essential that all three roles contribute to both the study process and the complexion of the ultimate study conclusion.

Any approach to the study of community health will likely generate more questions than solutions. To begin with, what do we mean by "community"? What is "health"? How do we define "health needs" and, once defined, how do we determine what is required to satisfy those needs? At first, these questions may appear purely academic. Yet the manner in which they are answered ultimately determines the efficacy, as well as the utility and validity, of the health care research and the effectiveness of the medical service programs that should emerge from community health studies.

These questions are often answered in one of several ways. The most usual way of determining community health status is through the study of morbidity and mortality data compiled by health departments. The city of Detroit, for example, is divided into 16 census tracts. Incidence rates for communicable diseases, accidents, suicides, homicides, births, and infant and maternal mortality are determined for each census area, and these data are compared between areas and relative to citywide rates to define health status. "Community" is represented by geographic census areas with no attention paid to population heterogeneity. "Health" is merely a matter of relative rather than absolute significance. Health needs are inferred rather than defined. Morbidity and mortality data are important to the understanding of health problems, but, by themselves, they are inadequate for the study and assessment of "true" community health.

What, then, is a systematic approach to the study of community health that will overcome these inadequacies? To begin with, the study implies the manipulation of identifiable variables in the pursuit of some conclusive statement about the subject under study. The conclusion may range from the recognition that no conclusion can be reached to the determination of specific characteristics of the study subject. Researchers should always be prepared to discover and admit to negative, positive, and even inconclusive results. All research experiences, regardless of their outcomes, are learning experiences and as such are productive.

The first step in the study of community health, therefore, is the definition of those variables bearing some relationship to the stated objectives of the study. Two types of variables form the basic content of a research study: independent variables which are, in the context of the study, nonmanipulative; and dependent variables whose characteristics are influenced by their relation to the independent variables. Many of these independent factors are the basic demographic characteristics that contribute to the life style of the study population. Factors such as ethnic or cultural affiliation, race, educational level, occupational skills, and income illustrate some variables defined as independent in terms of the study of community health. Dependent variables are those characteristic attitudes, values, and behaviors that are influenced by the existence of the independent variables. In this case, health needs, acceptance of the medical care system, and participation in health programs are factors determined by the "forces" of sociocultural, economic, and environmental factors.

When we speak of health care, we should recognize that the relevant community extends beyond the demographically-defined consumer population. Health care is a process based on the interaction of providers and consumers of health care services. As illustrated in Figure 1, our definition of "community" must therefore include the providers of care and the instrumentalities in which they operate. A proper study will define the independent and

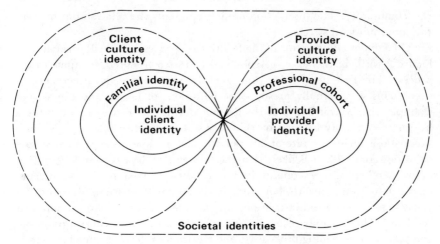

Figure 1. The health care community.
(From R. Meyer and M. Kolins, "Psycho-social Factors in Providing
Patient Services," in W. E. Moore (ed.), *Health Care Delivery—Crisis
or Challenge: Proceedings of the 18th Annual Stephen Wilson
Seminar* [Detroit: Wayne State University, 1971])

dependent variables characteristic of the providers, as well as the interaction
elements inherent in the health care process.

The supply of manpower, facilities, and money in the area under study
represent independent variables of the community's provider component.
Manpower includes physicians, dentists, nurses, allied health workers, social
workers, lawyers, and the many other human resources involved in provision
of health and welfare services. Facilities include all health-related physical
resources, ranging from neighborhood centers and school screening programs
to area hospitals, clinics, and private offices. Money, of course, reflects the
extent of insurance coverage and special benefit programs in the population,
as well as other sources of support such as the indirect payment for services
provided by health departments and the personal wealth of consumers.

The availability and accessibility of these resources are dependent
variables influenced by the supply. Such items as clinic hours, scope of
services provided, waiting periods, and other operational aspects of the health
care process are determined by the supply of resources in the community.
These factors also have a significant influence on the quality of care experienced by the community and the quality of the provider-client relationship
within that community.

This community of provider resources and consumers represents the
laboratory of the health care researcher. Different dimensions of the health
care process can be studied through manipulation of identifiable variables in

this community. The interaction model represents the relationship between the community subsystems.

The major components of the health care process can be illustrated as in Figure 2 with, in this case, psychosocial factors considered as inputs to the process. The first step in this process is the recognition and acceptance of need. This step actually presupposes definitions of health and illness. For example, the Detroit Board of Health may decide that census area K is not a healthy community because morbidity statistics indicate that area K has a tuberculosis incidence rate of 201 per 100,000 population as compared to a citywide rate of 72. It is likely that the residents of area K would agree that many members of their community suffer ill health. Despite this concurrence on the definition of health, however, the definition of health needs may differ vastly between the local institution, the providers of service, and community residents. The city may recognize "need" as a tuberculosis control program, complete with screening procedures and treatment referrals, aimed at detecting stricken persons and removing them from the community for therapy. The residents, on the other hand, may perceive "need" as a housing program that would provide less densely populated, well-ventilated homes, aimed at removing tuberculosis-inducing environmental factors from the community.

The health care researcher must seek out these different perspectives. If he fails to consider the many possible interpretations of need, his study of community health will reflect only part of the story and, as such, will result in distortion.

The acceptance of a recognized need also reflects the degree to which that need is perceived as relevant to other requirements of life. Providers and consumers in the community may have differing opinions on this matter. Each set of actors in a community has its own particular interests, and the acceptance of a recognized health need as deserving of attention reflects those interests. Given a chunk of money, therefore, providers would probably spend it on a program of medical intervention, while community residents might direct the money toward issues of higher personal priority, such as food and housing. Providers also have a set of priorities within health programs and, given limited supply, the contours of a proposed health care program may be unacceptable to those providers. In essence, they may feel that the need under study does not warrant the degree of intervention it is receiving.

Somehow health needs must be defined in terms of their severity in relation to the severity of other life-sustaining community problems. Once again, if the health care planner fails to consider relative priorities in the total community, the mode of intervention designed to tackle the recognized health need may be inadequate or inappropriate in light of other needs. Under-utilization and excessively high cost will be the outcomes.

Once a health need is recognized and accepted, the second step of the health care process develops. This step involves a client's seeking of health

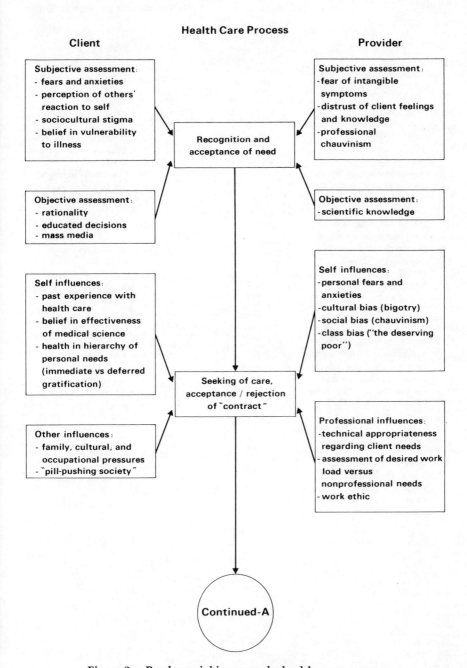

Figure 2. Psycho-social inputs to the health care process.
(From R. Meyer and M. Kolins, "Psycho-social Factors in Providing Patient Services," in W. E. Moore (ed.), *Health Care Delivery—Crisis or Challenge: Proceedings of the 18th Annual Stephen Wilson Seminar* [Detroit: Wayne State University, 1971])

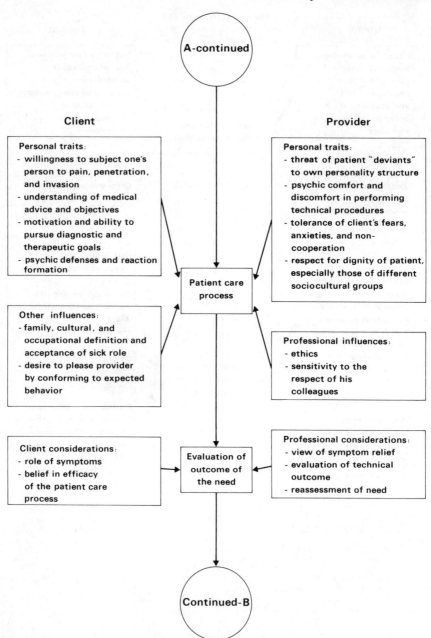

Figure 2 *(continued)*.

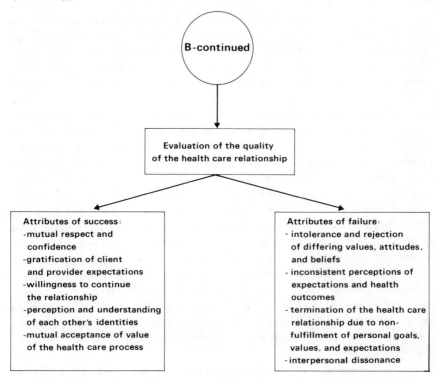

Figure 2 *(continued).*

care and his acceptance or rejection of the health service "contract." Traditionally, an area with poor health status—as defined by morbidity and mortality rates—has come to the attention of health care officials who then seek funds, facilities, and personnel to establish a health service program in that area. Many government programs have emerged in this manner. These programs have not always developed according to the community's definition of need; and, with limited consumer involvement, they will continue to be basically unresponsive to actual community problems. The greater this disparity between program organization and community desires and values, the more likely that the health program, despite its good intentions, will be unacceptable and unused by the clients it was designed to serve.

We are again in the realm of the quality of care and the health care relationship. This quality is reflected in all phases of the health care program's operation, ranging from the content of intake interview forms to the personal attitudes of health providers. Acceptability of a program can be assured only if mutual dignity, respect, and acceptance on both sides form the basic philosophy of the health service relationship. These characteristics are not always easy to identify and measure. Yet in the study of community health, the researcher must pay particular attention to these attitudinal and behavioral variables, for they often reflect an accurate account of community health care.

When an acceptable health care program in which both provider and consumer members of the community interact has been developed, the third phase of health care, the actual patient care process, is activated. Provider-client cooperation and mutual understanding is particularly important during this phase. It is here that the role of the sick and the service-professional role meet head-on, and the effectiveness of the patient care process in satisfying health needs depends on the compatibility of these two roles. Both provider and consumer behavior are related to the demographic characteristics of each group, because psychosocial, power, cultural, and economic attributes influence attitudes and behaviors. The growing emphasis on inclusion of community residents on health care teams reflects the need to integrate provider and client health-objectives; therefore, both parties actively strive for compatibility in their respective roles. The study of community health thus includes the study of health care organizations and the ways in which they tailor aspects of the patient care process to expressed community needs.

The last phase in the health care process is evaluation of outcome. Here the researcher must ask two basic questions: Has the service program influenced the community's health? and What was the quality of the health care process? The first question is simply asking if the original health need has been altered. To answer this, need must be reexamined in terms of both the providers' and the consumers' definitions. The process may have improved or diminished the community's health; or, if the mode of intervention was inappropriate to expressed needs, it may have made no impact on the state of community health.

Quality can be partially assessed by examining the structural relationships and attitudes of patients and providers involved in the health care process. If quality of care on either a personal and/or an institutional basis was unacceptable, the use of the health care program may decrease. The most important factor in studying the outcomes of health programs is the maintenance of communication between community members, both provider and consumer. These channels must be used for the exchange of complaints, advice, praise, and, in general, expression of valued needs.

The effective health care researcher and planner must approach the study of community health with understanding of both the social and medical sciences. His first step is to identify the independent and dependent variables of his study area, focusing on both providers and consumers in the total community. Tools for identifying these variables range from interview surveys and self-administered questionnaires to public health, clinical, screening, laboratory, and pathology records. In addition to seeking the signs of morbidity and mortality which result in disruption of daily activities, the researcher should seek an understanding of the attitudes, values, and beliefs of the total community regarding health and illness. New parameters for investigation are needed.

Assessment of the availability, accessibility, and type of resources is basic to an understanding of general dimensions of community health. For example, where there are no dentists or dental facilities, one cannot expect to find high levels of dental health; where there are severe money limitations, one cannot expect to find comprehensive health maintenance.

The health care researcher must also be sensitive to the quality of health programs as expressed by the acceptance of those programs by their potential clients. The community that is aware of its needs may also be aware of what will satisfy those needs. The rejection of a health program may, therefore, be a signal that community health problems have been misunderstood by health care planners.

Once the study variables have been identified, manipulated, and analyzed, the researcher must reach a conclusion about the health of the community. If his conclusion recommends some form of intervention, studied consideration must be given to what form would be most appropriate to the need. It may be that current programs are appropriate but inadequate. Intervention would then take the form of increasing the supply of resources in the community to facilitate program expansion. It may be that current programs are inappropriate; intervention would involve redesign. Inadequacies can be supplemented, unacceptable programs can be reformed, and deficiencies in communication can be eliminated through greater involvement of concerned community members and by education of the community in order to promote knowledgeable exchange of ideas and decision making in the pursuit of community health goals.

Once the form and level of intervention have been decided upon and implemented, the researcher must design a method for evaluating the outcome of those efforts and an appropriate data control system must be designed to compare delivery models. He must prepare himself to reexamine the dependent variables and to reassess the quality of individual and institutional care. The study of community health thus becomes a circular task which, through assessment, intervention, evaluation, and reassessment, becomes a continuous monitor of social change.

REFERENCES

Meyer, R., and Kolins, M. Psycho-social factors in providing patient services. In
W. E. Moore (ed.), *Health Care Delivery—Crisis or Challenge. Proceedings of the
18th Annual Stephen Wilson Seminar.* Detroit, Michigan: Wayne State University,
1971.

Murray, G. R. Systems analysis in health care delivery: Homeopathy or medicine?
The Pharos of Alpha Omega Alpha, 1971, *34*(1), 23-27.

CHAPTER 16

Systems Approach to Evaluation of Quality of Care in an Urban Hospital

Joyce Y. Passos

There are many paradoxes about research in a practice field, but the most impressive and most depressing paradox for the faculty member in a practice profession is that if one is interested in using the methods of research to improve practice, one must be involved in situational contexts in which practice is in fact taking place. The research part of that effort requires separate resources and separate time, while the resources available within institutions of higher learning for the practice professions have traditionally been more generously allocated to the teaching function—with the possible exception of medicine—and to a degree for the service function, rather than to the research component of the effort.

This paradox was important in our survey of an urban hospital's quality of care. Our experience was not unique; many other endeavors in the area of patient care research, specifically research in nursing care, have faced the same problem. Likely the problem will intensify.

Our survey used a systems theory approach. There are basically two systems methods, and any specific method is probably somewhere on the continuum—the hyphenated continuum—of inductive-deductive methodology. The empirico-intuitive method, although it stays very close to reality, is often criticized as lacking the elegance and strength of the more pure method of deduction. The difficulty in deduction is establishing whether the fundamental terms one is using are in fact correctly chosen (von Bertalanffy, 1968). One rarely uses only one or the other method; rather aspects of both are used, although the focus in exploratory or descriptive studies is perhaps closer to the empirico-intuitive method. This survey more nearly uses the empirico-intuitive method.

A number of questions have been raised about general systems theory and its application to phenomena of concern in the health field. One recurring question is whether or not general systems theory is essentially a physicalistic simile, perhaps inapplicable to psychological phenomena.

The second type of question asked is whether or not a general systems theory model has explanatory value when the pertinent variables cannot be defined quantitatively, as is generally the case with psychologic phenomena. Some of the answers given to these kinds of questions are encouraging. For instance, one type of answer is that the systems concept is abstract and general enough to be applied to any whole consisting of interacting components. That certainly is the case when we assume that, in the health professions, the person is the whole about which we are concerned and that he is more than the sum of his subsystems or parts.

The second type of answer that has been given is that if quantification is impossible, and even if components are ill defined, certain principles will qualitatively apply to the whole system. Explanation in principle, then, may be possible (von Bertalanffy, 1968).

It was for explanation in principle that we were striving in the survey of an urban hospital's quality of nursing care. General systems theory represents a complete revision of the original homeostasis principle which emphasized a tendency to equilibrium, or toward the discharge of tensions. There is an important argument for criticizing its use by the helping professions, because even in the area of medicine, which might conceivably be an exception, we are concerned about person development and not just about the maintenance of life-support systems of the body. In consideration of life-support systems alone, homeostasis or equilibrium is a viable concept. But in consideration of the development of *personhood*, equilibrium or homeostasis is probably not a very productive concept, because it violates many of the dimensions of human development which nursing and social work are concerned about. In other words, we may often want to dis-equilibrate people to get them to move.

In referring to Kant's maxim that experience without theory is blind, but theory without experience is a mere intellectual play, von Bertalanffy (1968) cautions that model building must not be a purpose in itself. Some graduate students — reacting to efforts to get them to consider model building as a first step in helping them to conceptualize the whole, whose parts they then are obligated to try to examine in some quantitative way — charge us with the sin of engaging them in an effort that looks to them like a purpose in itself. They often initially perceive this as an activity with somewhat limited value. But it is very easy, particularly when one is struggling to understand an area such as this, to get very much caught up with the semantic and conceptual enticement of the activity and to forget the purpose in the practice field for this activity.

Cybernetic models also present problems where one applies them to studies of health care. A cybernetic model obligates one to a certain kind of a system, a system that includes the feedback principle and imposes the obligation of automatic control systems, originally conceived of as the nervous system and brain, in relationship to mechanical-electrical communications systems—in other words, the machine-man interface. In the cybernetic model, controlled systems are governed and activated by feedback loops, which have inherent in them monitors, comparators, controllers, and regulators. One difficulty that arises when this model is applied to problems in health care is the fact that it cannot easily handle the domain of psychological phenomena. This is true even though cybernetic models are the neatest, cleanest, and the most satisfying when applied to the life-support kinds of behavior of living organisms. But with psychological phenomena, cybernetic models just do not hold up as well, perhaps because we do not know what needs to be monitored, what comparator mechanisms are possible to use, and what the regulatory functions are. Thus, although in the physiologic domain cybernetic models have served a valuable purpose in relationship to optimizing the support of life, I think they have tended to lead us astray in terms of getting the whole person into focus.

Definitions of "system" and "subsystem" have varied widely. Churchman (1968) gives an all-purpose definition, which has the advantage (or disadvantage) of using no mathematical or algebraic symbols. "A system is a set of parts coordinated to accomplish a set of goals" (p. 29). I believe the goal component is an important part of the definition. Unless one accepts "goal" as equivalent to "desired state" in the cybernetic model, the goal component is not inherent in that model. There is no reason why "goal" cannot be a legitimate part of the systems concept, however. Churchman stipulates that there are five questions to be asked about a system:

1. What are the system's objectives? What performance measures of the whole system can you propose? These are "state" variables, in relation to whatever level of phenomena you are defining as constituting the system.
2. What is the environment of the system? What are the fixed constraints, if you will, upon the system? These are the things over which the system has no influence.
3. What are the resources of the system? What are the factors within, and under the control of, the system itself?
4. What are the components, or the subsystems? In relationship to these components, what are their activities, their goals? What performance measures can you propose for the subsystem components?
5. How is the system managed? This is perhaps another way of asking what is in the box of the controller and regulator components within the cybernetic model (Churchman, 1968).

One strong stimulus to our undertaking of the Urban Hospital Survey involves the character of environment versus the system, a problem that has been addressed by several authors on the subject of general systems. They generally agree that environment, in some sense, is outside of the system's control and that factors in this environment determine, in part, how the system performs. The important aspect of the environment is that it provides the "requirement schedule," or the "givens," for the system (Churchman, 1968). An important problem here is that often, particularly in human systems, the "systems fail to perform properly simply because their managers have come to believe that some aspect of the world is outside of the system and not subject to any control" (p. 37).

Looking at the hospital as a social system and the delivery of nursing care as a subsystem, I would hypothesize that many of the problems experienced by professional practitioners derive from the fact that they perceive as environment those factors which are part of the professional service subsystem within the institution. I believe this is a critical misconception, not only in relationship to the nursing service subsystem within the hospital and other agencies but also for patients as a system within health care. One of the major objectives of all providers of health care should be to help the patient understand that many things which he perceives as environment are, in fact, a part of his system. It seems that our society is coming to appreciate this more and more, often to the great distress of health professionals.

Resources that are inside the system are the means to do the job,

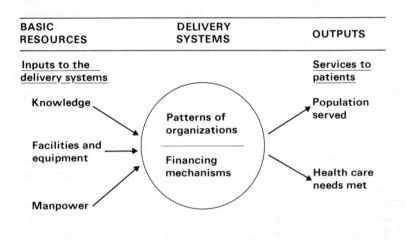

Figure 1.   Components of health care systems.
(Modification of figure by W. L. Kissick in "Health Policy Directions for the 1970's," *New England Journal of Medicine,* 1970, *282,* 1343-1354. Copyright Massachusetts Medical Society.)

whatever the mission of the system. The things the system can change and use to its own advantage constitute those resources (Figure 1).

One of the most serious limitations on the validity of efforts to evaluate the performance of delivery systems, including nursing service delivery systems, is that inputs have been used to measure outputs (Fein, 1967).

The Urban Hospital Survey attempted to evaluate the quality of the output of nursing service to inpatients and to measure selected aspects of the conditions under which that quality of care was being delivered. Some of those conditions of delivery were considered by the nursing service to be within its system, and some were considered to be in the environment and thus outside the control of the nursing service.

Hearn (1971) in conceptualizing social work as "boundary work" used a set of symbols to represent the known and unknown in the client system and its environment:

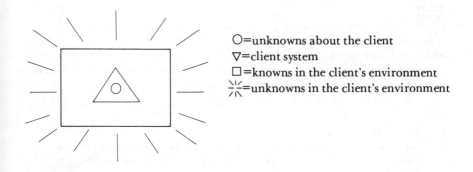

O=unknowns about the client
∇=client system
□=knowns in the client's environment
☼=unknowns in the client's environment

Using ∇ as patient, double solid lines for knowns in the patient's proximal environment, single solid lines for knowns in his intermediate environment, and broken lines for environmental unknowns, Figure 2 illustrates the regions within the nursing service system of an urban hospital from which variables were identified for study.

Convergence of three major circumstances led us to undertake the Urban Hospital Survey in the summer of 1970. The first was a change in approach to the teaching of medical-surgical nursing at the graduate level, which required assessment not only of patient-state variables but also of setting, or environmental, variables. This change in focus was based upon the premise that nursing has responsibility for processing data about system levels that are greater than the life-space boundary of the patient; that in order to do primary intervention, the nurse may have to operate on variables that are other than patient variables in order to achieve desired changes in the patient's state.

The second circumstance was the expressed need of the hospital involved

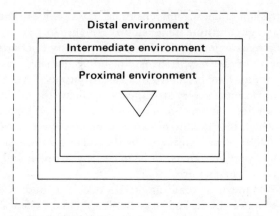

Figure 2. **Regions within the nursing service system of an urban hospital from which variables were identified for study.**

= Dependent variable, quality of nursing care provided to inpatients
The tool used was a quality of patient care scale (Wandelt, 1970), in which scores represent a judgment of quality of interaction between patient and each person in his proximal environment, as judged by a professional nurse observer during a two-hour nonparticipant observation of all interactions with a patient.

= Independent variables, conditions of delivery.
These are factors within the proximal and intermediate environments believed to influence the quality of nursing care provided to inpatients in an urban hospital.
Categories of conditions of delivery:
1. Selected patient characteristics
2. Medical components
3. Ward and hospital management components

for baseline data on the quality and conditions of nursing care being provided to inpatients, prior to some major changes proposed within the nursing service department.

The final reason was the availability of a tool to measure quality of nursing care provided to hospitalized patients.

The tool used to measure the dependent variable, quality of nursing care provided to inpatients, was the Quality Patient Care Scale (QualPaCS), a 68-item scale for rating the quality of nursing care (Wandelt, 1970). Ratings depend upon the clinical judgment of nonparticipant observers who are professional nurses. In the Urban Hospital Survey, the qualifications for observers included a master's degree in nursing and a minimum of five years' experience. Professional nurse observers complete a nursing assessment on each patient selected in a stratified random sample of the inpatient population and then observe and rate all interactions between the patient and nursing personnel during a two-hour period. Ratings are distributed among the 68 items of the QualPaCS, which are grouped into six areas of patient need:

	Area	Number of items
1.	*Psychosocial: Individual* — actions directed toward meeting psychosocial needs of individual patients	15
2.	*Psychosocial: Group* — actions directed toward meeting psychosocial needs of patients as members of a group	8
3.	*Physical* — actions directed toward meeting physical needs of patients	15
4.	*General* — actions that may be directed toward meeting either psychosocial or physical needs of patient or both at once	15
5.	*Communication* — communication on behalf of patients	8
6.	*Professional implications* — care given to patients reflects initiative and responsibility indicative of professional expectations	7
	TOTAL	68

The Urban Hospital Survey was based on five assumptions: (1) that the QualPaCS is a valid measure of quality of nursing care; (2) that the scale provides for scoring items on an interval scale from 0.0 to 5.0 (for this study it is assumed that the midpoint of that scale, 2.50, represents average care); (3) that the demonstrated level of inter-rater reliability among professional nurses trained in using the QualPaCS ($r=0.73$) is acceptable for the purposes of this study; (4) that the quality of care received by a stratified random sample of patients on a single ward will be representative of care received by all patients on that ward; and (5) that the effect of the presence of a non-participant observer on the quality of nursing care received by patients cannot be measured; however, the assumption is that whatever effect the observer's presence has, it will be comparable across wards.

We posed two groups of focal questions, the first about the quality of nursing care given to inpatients, and the second about the factors that might influence the delivery of nursing care to inpatients.

The first group of questions related to a measure of quality that would characterize nursing care provided throughout the hospital, to the range of variability in quality of nursing care across wards, and to comparisons of quality of nursing care according to such factors as weekdays versus weekends, shifts, and areas of patient need identified in QualPaCS. The second group of questions related to comparisons of quality of nursing care according to such factors as type and number of medical services on a ward, seriousness of illness of all patients on a ward, level of nursing care requirements of all patients on a ward, use of written nursing care plans, type and number of nursing personnel on duty on a ward, and level of medications activity.

In relation to the first group of questions, we found that the mean quality of nursing care for all wards was considerably higher than we hypothesized; the range of variability across wards was as great as we hypothesized; there were differences in quality of nursing care according to shift, but not in the direction we predicted; quality of nursing care was higher Monday through Friday than it was on weekends; and the area of need for which quality was consistently the lowest was the area we had predicted would be lowest.

We found that: (1) there were differences in quality of nursing care on medical and surgical wards but not in the direction we predicted; (2) there were differences in quality of nursing care according to number of medical services on the ward; (3) quality of nursing care appeared to vary, as we predicted, with degree of illness and level of nursing care requirements of patients; (4) presence of written plans for nursing care seemed to be positively correlated with quality of nursing care; (5) there may be a critical ratio of professional to nonprofessional nursing personnel beyond which quality of care is jeopardized; and (6) analysis of medications activity on all wards within the hospital was welcomed by all levels of personnel, in and out of the nursing department, and this aspect of patient care seemed to represent an ideal talking point for bringing together levels of personnel within nursing as well as various departments within the hospital.

REFERENCES

Bertalanffy, L. von. *General Systems Theory: Foundations, Development, Applications*. New York: Braziller, 1968.

Churchman, C. W. *The Systems Approach*. New York: Dell, 1968.

Fein, R. *The Doctor Shortage: An Economic Diagnosis*. Washington, D. C.: Brookings Institution, 1967.

Hearn, G. The client as the focal subsystem. Paper presented at the Health Research and the Systems Approach Conference, Wayne State University, Detroit, March, 1971.

Kissick, W. L. Health-policy directions for the 1970's. *New England Journal of Medicine*, 1970, *282*, 1343-1354.

Wandelt, M. A. *Quality Patient Care Scale*. Detroit: College of Nursing, Wayne State University, 1970.

CHAPTER 17

Systems Approach to Studying Continuing Health Professions Education Needs

Joseph W. Hess

The traditional approach to continuing education has been a simple one: one or several "experts" decide what should be taught and the learners are identified on the basis of personal interest or their own perception of need. Very little attention is given to the more fundamental question of educational need determined on the basis of the care actually being given to patients as compared with a defined standard of care.

If we accept the view that patient care is the most important output of the health services system, then at least part of the basis for determining the educational needs of practicing health professionals should be the quality of care their patients receive. If one takes a systems approach to identifying educational needs, then it is useful to define continuing education as any planned intervention that beneficially alters the patient output of the health services system.

I believe the two foregoing premises are valid educational concepts for the continuing health professions education program. Stated in a slightly different way, these premises are: (1) the output of the health care system should be an important factor in determining educational need, and (2) education is any type of planned activity or intervention that improves the patient care output.

The model for continuing education presented here is much more complex than the traditional model. Credit for elaborating the basic concepts incorporated in the continuing education program to be described belongs especially to Williamson, Alexander, and Miller (1967), and Brown (1969).

The types of units that are participating in this continuing education program are shown in Figure 1. Four hospitals are linked together through a continuing education project staff consisting of a part-time physician serving as coordinator, a part-time project director, and a part-time secretary. An advisory committee, with representatives from the fields of medicine, nursing, education, and behavioral science, aids the program staff in guiding and implementing the subject. Through the assistance of the continuing education project committee, the resources of Wayne State University are available to the participating hospitals to assist them in accomplishing their principal goal of improving patient care in their own institutions. Three of the four hospitals are small inner-city hospitals of approximately one hundred beds, none of which has previously had a formal type of affiliation with the University. The fourth is a large public hospital affiliated with the Wayne State University Medical School.

Each of the hospitals is an autonomous unit with its own board of directors, professional staff organization, and professional staff membership. Some of the medical staff are members of more than one hospital staff and other informal relationships exist, but the major formal linkages between these hospitals comes through this continuing education project.

Each hospital has its own unique internal organization for participation in this project, but there are common features, as outlined in idealized form in Figure 2. Each hospital has a physician coordinator for continuing education who is the liaison person and project supervisor. He is assisted by a full-time health records analyst who works under his supervision. This hospital team of physician coordinator and health records analyst is the only new organizational element that has been introduced into the hospital system. The coordinator provides information-gathering and liaison functions between the patient care, medical records, or audit committee and the committee concerned with ongoing professional education. These two committees, in turn, are responsible to the executive committee of the hospital staff. The decisions made by the hospital executive committee and the continuing education committee influence the performance of the professional staffs, which in turn should influence the patient care system. As indicated by the box in the lower right-hand corner of Figure 2, part of the coordinator's role is to assist in making needed external educational resources accessible to the hospital to beneficially alter the performance of the patient care system.

The arrow in the lower center portion of the diagram pinpoints a fundamental function of the coordinator-health records analyst team, that is, to monitor the performance of the patient care system as reflected in the health care records of the hospital.

A variety of sources can be used for making initial hypotheses concerning education needs within a given institution. National studies have led to the

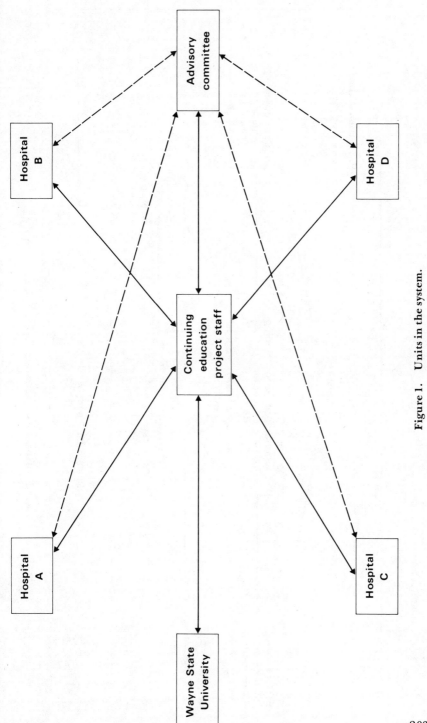

Figure 1. Units in the system.

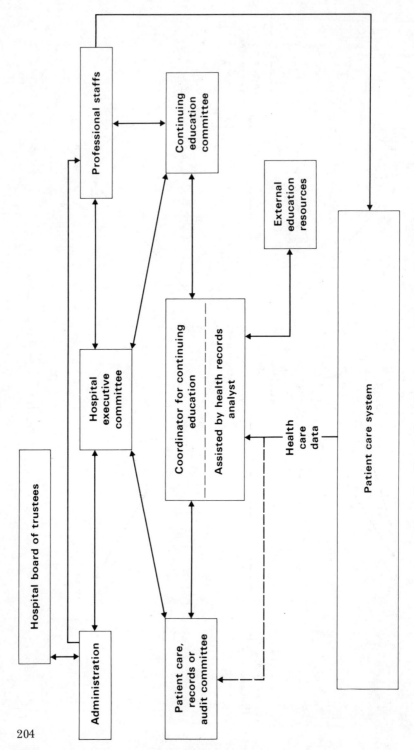

Figure 2.   Hospital organization for continuing education.

setting of national priorities for health care. One of these was the study that resulted in the Regional Medical Program (RMP) legislation. This national assessment of need identified heart, cancer, and stroke and related diseases as major national causes of morbidity and mortality. National and regional surveys have also provided demographic data for identifying populations at greatest risk for various types of illness. Likewise, community studies can provide some guidance in identifying the special needs of various segments of the community. The judgment of the professional staff working within the hospital is another important element in defining the areas of patient care that should be studied to determine what educational needs are the most pertinent for a given institution. Step one, therefore, is an amalgamation of data and judgment from a variety of sources to formulate an initial educational hypothesis (Figure 3). The next three steps are designed to test the initial hypothesis. Step two, in the flow diagram, is a definition of acceptable criteria for care related to the hypothesis. This is done by the patient care or audit committee with the guidance and assistance of the physician coordinator and the continuing education project staff, the advisory committee, and other resources as needed.

When the criteria are defined, the hospital team (physician coordinator and health records analyst) collects patient care data from a random sample of patient's records (Step 3). The data are then analyzed and organized into a form that will enable the committee to determine whether or not care meets the criteria. If care does meet the criteria, then the original hypothesis is rejected and another problem is selected for study, and the cycle begins again (Step 5). If, on the other hand, care does not meet the established criteria, the committee formulates hypotheses for effective corrective action (Step 6). These hypotheses are then translated into educational programs designed to correct the identified deficiencies (Step 7). After the programs have been put into operation, the health team again collects data and determines the effectiveness of the education intervention. If the program now meets criteria, the patient care and education committees can turn their attention to selecting other problems for study. If performance does not meet criteria, it is obvious that additional effort and perhaps new approaches must be developed to bring the performance of the system to an acceptable level.

The program described above has just reached stage 4, with the first patient care problem being carried through this process. The patient care problem selected for study was the identification and management of patients with hypertension. Figure 4 provides some examples of the kind of data that are coming out of the study.

A random selection of 575 charts was made from the population of patients discharged from one hospital during the first six months of 1970. Of those 575 patients, 112 had a diastolic blood pressure greater than 90 mm/Hg on admission. The blood pressure of 35 of these patients returned to normal by the third recording and they were eliminated from further consideration.

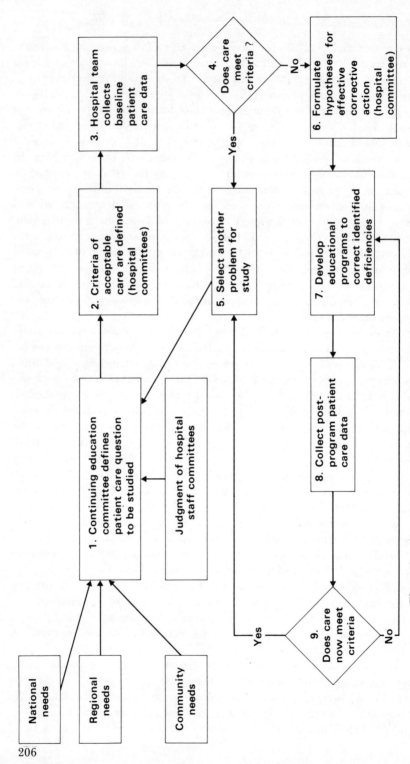

Figure 3. A systems approach to studying continuing education needs for health professionals.

National needs

Regional needs

Community needs

1. Continuing education committee defines patient care question to be studied

Judgment of hospital staff committees

2. Criteria of acceptable care are defined (hospital committees)

3. Hospital team collects baseline patient care data

4. Does care meet criteria ?

No

Yes

5. Select another problem for study

6. Formulate hypotheses for effective corrective action (hospital committee)

7. Develop educational programs to correct identified deficiencies

8. Collect post-program patient care data

9. Does care now meet criteria

Yes

No

206

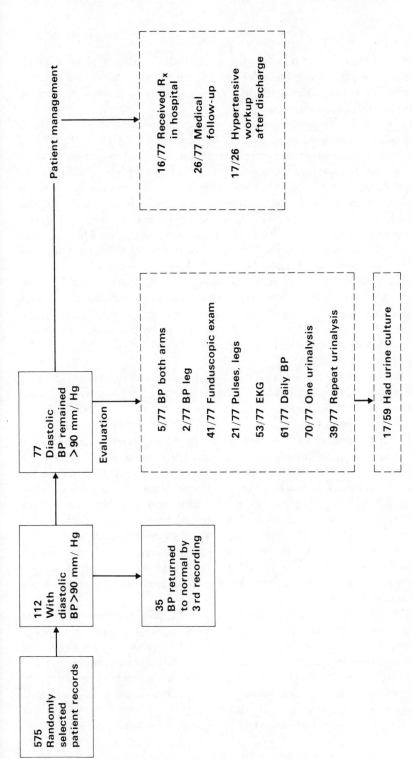

Figure 4. Sample patient care data.

The primary study population consisted of 77 patients whose diastolic blood pressure remained above 90 mm/Hg on three consecutive recordings following admission. Only a few of the criteria for evaluation of the hypertensive patient are shown here.

The hospital committee said that, among other things, all patients whose blood pressure remained above 90 mm/Hg on at least three recordings should have blood pressure recorded in both arms and one leg, should have a funduscopic examination, and should have recording of the quality of the arterial pulses in both legs, an EKG, daily blood pressure, and at least one urinalysis. Criteria were also established for determining which patients should have a urine culture. The figures in the evaluation block of Figure 4 indicate the degree to which each of these criteria were met. There were 59 patients who, according to the hospital committee, should have had urine cultures. Cultures were obtained for only 17 of these patients.

The committee in this hospital stated that the patients should either receive treatment for hypertension in the hospital or provision should be made for medical followup of their hypertensive problem. Furthermore, if the patient did not have a hypertensive workup in the hospital, then this should be done after discharge. The data in the right-hand box indicate the extent to which these criteria were met.

The data presented here tell only part of the story. A much more detailed analysis of selected records must be carried out in order to answer the many questions that the data raise.

The main point to be made, however, is that none of the deficiencies in patient care identified in this study are likely to be affected by conventional continuing education programs whose primary purpose is to increase the knowledge of individual health professionals. Without question, every physician on the staff knows how to take a blood pressure in both arms, how to do a funduscopic examination, and how to order urine cultures. The majority probably also know how to treat hypertension, how to provide for medical follow-up of patients with hypertension, and how to either do or request a hypertensive workup after discharge from the hospital. Rather, the problem for the professional staff is to agree upon how and when knowledge that they already possess should be applied to the patients under their care, and then act accordingly.

This model of continuing education can be applied to any type of patient care problem involving one or several types of health workers. The types of problems studied can be simple or complex. The type of data collected need not be confined to that recorded in patient's charts. Direct observation and data collection by a variety of methods are fully compatible with the model.

The advantages of this model over traditional models of continuing education include these two important features: assumptions concerning educational need can be tested, and the impact of educational intervention on health care is assessed as an integrated part of the process. Still to be studied is a comparison of the cost-benefit ratio of this approach versus the traditional model.

The model described in this paper may be regarded as a more scientific approach to the process of continuing education in the health professions than that which has been used traditionally. It includes the broad perspective of the systems approach by establishing goals and criteria in terms of the patient care output of the health services system. The model provides for formulation of hypotheses, both in terms of educational need and for educational interventions. These hypotheses are not left as untested assumptions; rather, data on health care measurements are collected, and these can be used to confirm or reject the hypotheses.

This model of continuing education adds a very important systematic feedback loop to a health care system as defined by a hospital, something which is lacking in nearly all contemporary health care institutions. In so doing, it imitates the homeostatic mechanisms that are integral parts of most, if not all, effectively functioning living systems.

REFERENCES

Brown, C. R. Relationship of patient care to continuing education. Paper presented at the Eighth Annual Conference on Continuing Medical Education, Association of American Medical Colleges, Cincinnati, October, 1969.

Williamson, J. W., Alexander, M., and Miller, G. E. Continuing education and patient care research. *Journal of the American Medical Association*, 1967, *201*, 938-942.

CHAPTER 18

The World of Work-The World of Client: A General Systems Approach to Physiological Nursing Research

June C. Abbey

To discuss the use of a systems approach to study physiologic controls, one is forced to present the overall conceptual framework out of which the researchable problem grew. Belief in the aims of general systems, such as a need for interdisciplinary theories, order out of chaos, maintenance in a world of change, or purposiveness, self-differentiation with increasing complexity (von Bertalanffy, 1968; Rapoport, 1968) does not necessarily imply complete understanding or even complete agreement with every definition. Often, when relating a construct to a specific profession, the need arises to extend and even modify some of the concepts. This action forces a closer study of the theory, particularly in the area of possible obscurities. It demands a statement of position and clarification of definitions. While holding all of the inherent power of the atom, a theory is, after all, only a construct to facilitate greater understanding. Preferably, in nursing, it is a set of principles that are of value to clinical care and research.

By definition, a system consists of component parts that mutually react, which are held in place, as it were, by the exchange between them. The positioning is determined by energy, either of a positive or negative attraction or repulsion. The system thus develops boundaries (Hearn, 1958) and acts as a whole. A dysfunction of any part causes a system disturbance, not simply the loss of a single function (Abbey, 1970).

Energy, therefore, is the force that holds the system together and determines its limitations or boundaries. Introduction of energy into the system or an external force to the perimeter can yield repositioning, new alignments, and thereby change the configuration. To maintain itself, the system develops servodynamic mechanisms to offset and compensate for these strains.

To function, to be spontaneously active, to differentiate, and to evolve to a higher order of complexity, the system transduces energy into other forms, such as heat, movement, chemical reaction, electrical potential, or communication. In transduction, or change of form, some energy is lost to entropy. Energy is also lost as entropy in the struggle between components to maintain relationships contributing to the total functioning of the system. If this lost energy is not replaced, the system will contract.

While appreciating the usual differentiation between closed and open systems in which the open system can exchange with its environs, a more useful format for the study of physiologic controls defines the closed system as one in which activity or function has a high order of predictability, thus allowing for manipulation of a supposedly closed system, even if through the enveloping, surrounding open system.

The developmental organization of a system apparently has three aspects: identification, servodynamic mechanisms, and affirmation. These relate to the overall system so logically that they are the determinants of the "self of systems" (Abbey, 1970). Identification of self occurs when the system begins to act as a whole. This is a formative state in which components select position, in which attractions and repulsions are strong and entropic loss high, and in which the system is very unstable. Essential servodynamic mechanisms of self form and these feedback loops maintain the steady state of the system. In addition they have "the property to adjust future conduct by past performance" (Wiener, 1954, p. 23)." As sensors, these systems stimulate, facilitate, disseminate, and inhibit within the overall self. Important to physiologic nursing research is their ability to reset the reception level.

With the development of the servodynamic mechanisms, the self system gains stability. Strains are tolerated better because adjustment is possible. Of equal import to clinical research is that the feedback loops are subject to potential conceptual closure as prediction regarding their activities becomes more accurate. With the ability to withstand strain, the system begins to affirm self through increased stability, extension, and, in living organisms, increased relationships with other systems.

The foregoing overall format is central to the research design used in the following briefly described clinical nursing studies of physiologic controls. Because of the difficulties involved in obtaining instruments that guarantee complete safety for patients, the pilot studies are only just now under way.

The two projects relate to nursing management of patients with temperature regulation problems: shivering while on hypothermia blankets. and control of hyperpyrexia. The important aspects of general systems theory are as follows: (1) open systems can receive or discharge energy to their surroundings; (2) closed systems have a high order of predictability; (3) servodynamic mechanisms of self can stimulate, facilitate, disseminate, and inhibit; (4) sensors (servodynamic mechanisms) can reset sensing level; (5) the system is unstable while identifying self; and (6) affirmation of self accompanies increased stability.

The patient who is on the hypothermia blanket, whatever his diagnosis, is in a critical, low-energy systems state. Originally the study was designed to use postoperative open-heart surgical patients as subjects. The dangers of shivering with its concomitant rise in metabolism, and an increase in heart rate and stroke volume, coupled with drift and instability of body core temperature, all represented hazards to the patient's physiologic well-being and the need for nurses to use control measures to offset the use of drugs. A major difficulty was the presence of a shift away from hypothermia for the cardiosurgical patient; consequently, patients with increased intracranial pressure who were placed on hypothermia were substituted as the patient population.

A servodynamic mechanism pertinent to the study is the fact that shivering begins in the masseter and pectoral muscles. The beginning contractions, although largely invisible, can be measured by electromyography at inception. The predominant sensors for shivering are located in the four extremities; therefore, they are available to application of heat. Core temperature can be assessed by either a turbinate thermistor or esophageal thermistors. Readouts can be attached for simultaneous recording of thermistors and electrodes. If the extremities are warmed, shivering will stop. The temperature applied to the extremities can be closely controlled by having skin thermistors with set cutoff temperatures attached to the heating device. Thus, the first three of the aforementioned aspects of general systems theory were utilized in the design—an open system received energy as heat, the closed systems of extremity sensing had a predictable result, and shivering was inhibited by the servodynamic mechanism.

The second study involves youngsters between two and six years of age who have temperatures of 103°F or greater and who can be immersed in warm water. In this study the great problem, in addition to the physiologic assault of simple temperature elevation, is the oscillation of core temperature that follows the standard nursing measures for decreasing fever. The fluctuations of temperature can vary as much as four degrees in waves of chilling and fever after sponge baths or the application of ice bags to large superficial vessels.

In the design for this study, the primary factor of general systems theory used is the ability of an open system, the patient, to lose excessive heat energy to the surroundings. The practice of placing the child in a tub of water at

100°F is not new. This water temperature and the child's core temperature can be monitored by thermistors with correlative readout. The water temperature can be maintained at 100°F by adding cool water as this temperature increases from the heat absorbed from the child. Keeping the child at 100°F for 30 minutes will allow sensors to reset to this level. Instability (and fluctuation) of the system will be avoided as the system reidentifies self. At this lower temperature the system will begin to affirm self as the child feels better.

Here, then, are studies related equally to the world of work and the world of the client. They furnish a model for modification in care and in application of patient care designed to promote the identification of self in each, servodynamic mechanisms in both, and an opportunity for affirmation of self systems for better articulation in this dyad of the health services.

REFERENCES

Abbey, J. C. A general systems approach to nursing. In J. Smith (ed.), *Improvement of Curricula in Schools of Nursing*. Boulder, Colo.: Western Interstate Commission for Higher education, 1970.

Bertalanffy, L. von. General systems theory: A critical review. In W. Buckley (ed.), *Modern Systems Research for the Behavioral Scientist*. Chicago: Aldine, 1968.

Hearn, G. *Theory Building in Social Work*. Toronto, Canada: University of Toronto Press, 1958.

Rapoport, A. Foreward. In W. Buckley (ed.), *Modern Systems Research for the Behavioral Scientist*. Chicago: Aldine, 1968.

Wiener, N. *Human Use of Human beings*. (2d ed.) New York: Doubleday, 1954.

CHAPTER 19

Theories "Produce" Findings: Studying Effects of Innovation in Hospital Nursing Units

Shirley Smoyak

Theory is a powerful instrument. It is capable of changing the shape, scope, and character of what one subsequently observes. Depending on what theory one chooses as a guiding framework, the outcome in terms of description offered, explanation outlined, or prediction made will vary greatly. The methodology used produces the kinds of findings the researcher initially had in mind. That is not to say that the serious social scientist consciously presents biased data, but his preconceptions of what reality is, or how one should best proceed to explore some facet of a social situation, do in fact change what is observed.

This paper will show how differing theories of social systems produce different findings. The social system being considered is the community general hospital, and the specific situation is what happens whan a rational, on-line, real-time computer is installed as the mechanism for a hospital information system.

THE HOSPITAL AS A SOCIAL SYSTEM

As noted by Georgopoulos and Mann (1962), the community general hospital could easily claim the dubious honor of being one of the least researched of modern large-scale organizations:

In spite of its crucial function of aiding the integration and stability of society, through the maintenance of a level of health that permits other social institutions to accomplish their objectives, and in spite of its far-reaching impact upon nearly every facet of everyday life — particularly our economy, standards of living and community welfare — the community general hospital has not received more than a fraction of the scientific attention that its importance as an organization would warrant. As yet, our understanding of its functioning, problems, and characteristics is extremely limited (p. 588).

What have been studied in the field of health are the areas of diagnosis, treatment, prevention, and epidemiological aspects of disease. There have been a few largely descriptive studies of the career patterns and occupational dilemmas of health professionals. The psychiatric hospital has probably been studied more than any other type, but the study of any hospital as a complex social system is in its infancy. Further, recent investigations claiming to study the hospital as a "system" in fact use the traditional logic of cause-effect analysis or state emphatically that the study is descriptive and exploratory only.

Throughout the lengthy report on their comparative hospital study, Georgopoulos and Mann (1962) make reference to the hospital as a "complex organization" or a "complex social system." Although system theory may have served as their guiding framework, exactly how it is implemented in the empirical study is not clear. They apparently see the total hospital, and, to some degree, the individual departments, as the system to be studied. Their data, however, consist of responses to traditional questionnaires and interviews. Interdepartmental "interactions" are studied by analyzing the collective responses of individuals in the separate departments. They use instruments to study pieces or parts of the system. But a system is more than the sum of its parts. There is a need for systems theory and methods of study to be applied to the structure and dynamics of a complex system as a whole.

Probably the most significant factor contributing to the lack of systems theory and method applied to the study of hospital systems is that such theory and method have yet to undergo the translation necessary for its implementation in sociological research. As Buckley (1967) notes, "sociology, however, has yet to feel the impact of modern systems research. The little that has been written of sociological relevance by non-sociologists has not, for the most part, been presented in terms directly applicable to the field" (p. 55). Whereas Buckley is most useful in providing a clear description of the direction in which systems thinking and researching should go, he stops short of providing explicit methodological help:

We shall not take up here the question of exact specification and measurement of degrees of systemness, organization, or "entitativity." Current statistical techniques of association, correlation and variance analysis, and the like, can be considered, of course, crude measures of these. However, the problem of specifying and measuring the degree of organization or structure seems solvable in principle by such techniques as those of information theory and graph theory, though the era of their application to social organization has barely begun (p. 45).

Another major hospital study, currently in progress, furnishes an ideal situation for applying systems thinking. Bernard Goldstein and Coralie Farlee are co-directors of this project at Rutgers University. Whereas the Georgopoulos data are largely supplied by individuals about their own orientations, attitudes, and values and are collected into composite or collective measures in the analysis, Goldstein and Farlee are including analysis of nursing units as one of their methods, in addition to using questionnaires, interviews, and observation (both participant and spectator varieties), computer output, and other hospital records.

THE HOSPITAL INFORMATION SYSTEM (HIS) STUDY

The administrators of a 600-bed general, short-stay hospital in a medium-sized city on the East Coast decided to introduce a computerized hospital information system, using the on-line, real-time format. They invited a group of sociologists from Rutgers to study the process and implications of the technological change that would result. While the hospital is not associated with a university, its main functions include sponsoring teaching and research programs, as well as rendering service.

The director of nursing, although she viewed the innovation favorably, was not a part of the original decision-making body. She was brought into the planning and implementation phases after initial arrangements with the vendor had been made. Members of the nursing department below the level of director, were not consulted in advance about the philosophy or objectives of the system, nor have they participated to any extent by explicit invitation in its subsequent evolution (Goldstein and Farlee, 1971).

From the literature, and from their interviews with vendors, users, and other informants, Goldstein and Farlee have abstracted six major goals upon which a philosophy of implementation and operation of computerized hospital information systems might be formed. These are:

1. Increased speed and accuracy in the transmission of information,
2. More efficient scheduling of hospital personnel and resources, and of patients,
3. Better cost accounting and increased recovery of charges,
4. Improved patient care and safety, including better records of patient treatments and medications,
5. Improved research programs, utilizing patient records as a data base, with a focus on tests, diagnosis, and medications,
6. Better educational programs for interns, residents, and other hospital personnel (Farlee and Goldstein, 1970).

The apparent philosophy of the administrators and medical staff of this particular hospital has been a combination of more rapid dissemination of information and better record-keeping. However, at various times, other hospital personnel have mentioned "cost accounting" and "better patient care" aspects of the philosophy (Farlee and Goldstein, 1970).

Since the decision to introduce this change into the hospital was a unilateral one, from the top down, one wonders whether the administrators thought the change would be accepted. Judging from my own conversations with the chief administrator and his assistant assigned to the computer project, I believe that their view of the nature of man can be summarized as follows: Personnel here will accept this change because it is the rational thing to do. Furthermore, they will accept it with a minimum of information about it. Administration knows what is right and is in the best position to make such decisions. In Chin and Benne's schema (1969), they would be categorized as holding the empirical-rational approach. Within this schema, their view could not be described, on the one hand, as being very concerned with any personal relearning of the staff, nor were they really very coercive, on the other hand.

The project began in October 1967. Since then a variety of observational and analytical staff has been at work under Goldstein's direction. A number of reports and papers have been made and presented. These serve as the basis for some discussion here of findings.

THE SYSTEMS PERSPECTIVE IN THE STUDY OF INNOVATION

Buckley (1967) provides an excellent review of the dimensions of modern systems theory that are useful in sociology. He specifies the major premises and principles, giving examples from emerging social theory. Whereas it is quite clear that tremendous benefits can accrue if the focus used is studying the "complex of elements in mutual interaction," it is much less clear just how one goes about doing this. With only incomplete methodological equipment available for the study of systems, such an attempt is something like trying to take movies with a slide camera. Computers, because of their great speed and storage capacities, might at first appear to furnish the answer to the dilemma; however, they are actually linear instruments, whereas system data are essentially nonlinear. Thus computer simulation strategies and games are fated to remain just that — games, and not sufficiently informative of live processes.

The outlook is not totally bleak, however, and there have been some scattered improvements in methodology. Some of these are cautions, or consist of advice of the "how-not-to" variety. Buckley (1967) summarizes Campbell's argument that one should not assume axiomatically that social groups being studied constitute entities or systems. Rather than assuming that a system exists, one should formulate and test a hypothesis about the systemness of the social group. For instance, it might at first seem reasonable to assume that a nursing unit or a hospital department constitutes a system, or that a group of hospital workers brought together to serve one particular patient is a system. The danger in such an assumption is that a general psychological problem of how we perceive entities exists and overpowers reasoning. Campbell (1958) wrote:

The natural knowledge processes with which we are biologically endowed somehow make objects like stones and teacups much more "real" than social groups or neutrinos, so that we are offended by the use of the same term "real" to cover both instances. . . . Two sources of the feeling can be suggested. First . . . groups are in most instances less "solid," less multiply confirmed, of less sharp boundaries, less "hard." Second, and most important, we have evolved in an environment in which the identification of certain middle-sized entities was both useful and anatomically possible. As a product of this evolutionary process we have the marvelously effective mechanism of vision, which, within a limited range of entities, analyzes entitativity so rapidly and vividly that all other inferential processes seem in contrast indirect, ponderous, and undependable. More important, the visual process is so powerful and seemingly direct that we usually do not stop to notice the inferential steps involved in it and the nature of the clues employed (p. 17).

What follows is a partial listing of criteria of entitativity which lead discrete elements to be perceived as parts of a whole. These elements include proximity, similarity, common fate, and "pregnance" or completed boundary (Campbell, 1958).

Other possibilities for handling nonlinear relations include such strategies as step functions, buffer mechanisms, and primacy of variables. A step function is a special kind of nonlinear relation referred to when a variable has no appreciable effect on others until its value has increased or decreased by some minimal increment. Unless this is considered, research may fail to show any significant relationship, even though a large potential interaction exists. Buffer mechanisms differ somewhat in that they delay the effects of a variable until a later point. The primacy factor cautions against assuming that contributing variables have equal weight (Buckley, 1967).

Theoretical models of social systems include organic, mechanical, and process varieties. The first two are less useful than the process type in studying social phenomena. The basic question posed by the process model is: "How do interacting personalities and groups define, assess, interpret, *verstehen,* and act on the situation?" (Buckley, 1967, p. 23).

EQUIFINALITY VERSUS THE PRINCIPLE
OF CAUSALITY

In their analysis of the various nursing units in terms of their adaptations to the HIS innovation, the study teams' underlying principle was quite like the classic principle of causality. This principle holds that similar conditions produce similar effects and, consequently, dissimilar results produce dissimilar conditions. "Conditions" are specified as size of units, types of services rendered on the units, kind and quality of the personnel, and so on. Goldstein and Farlee (1970), for instance, made an effort to categorize the nursing units according to the degree of "medicalness" or "surgicalness." They pointed out that it is difficult to know for certain the referent for the terms "medical unit" or "surgical unit" as usually used in the sociological literature. Therefore, they proposed using a scheme whereby the unit would be referred to in terms of the percentage of patients who actually have a surgical diagnosis and ranked units by the percentage of hospital days generated by surgical cases. They then studied two measures of accommodation or adaptiveness as dependent variables.

The first accommodation measure focused on the discharge process, which normally occurred on the day shift (7 A.M. to 3:30 P.M.). The HIS system was programmed so that when a patient was ready to leave the hospital, a "discharge complete" entry was to be made, usually by the unit nurse or delegate, to the central pool. The relevant pieces of information in the entry included the time (from a computer clock) at which the message was, entered and the time of the discharge as supplied by the person phoning in the message. "Since the two times should be fairly close, but in practice are not, it can be argued that the lower the elapsed time, the greater the adaptiveness" (p. 12). Goldstein and Farlee referred to this measure as Adaptiveness I.

The second accommodation measure focused on the census report, which normally was printed out at the nursing unit about 11 P.M. In the system as programmed, personnel of the nursing unit were required to verify the census within an hour. The time that elapsed between the printing of the census and the verifying phone call was called Adaptiveness II.

Then, using the work of Hage and Aiken (1970) as the organizing framework, Goldstein and Farlee asked whether adaptiveness is positively associated with specialization and employee satisfaction, and whether it is negatively associated with centralization, formalization, stratification, productivity, and efficiency. Their chief question was whether surgical units are different from others from the standpoint of the variables specified above.

This classical principle-of-causality approach may present serious difficulties, however, if one's major goal is to understand the social system. "Bertalanffy, in analyzing the self-regulating, or morphostatic, features of open biological systems, loosened this classical conception by introducing the concept of 'equifinality' " (Buckley, 1967, p. 60). The principle here is that a final state may be reached by any number of devious developmental routes. Dissimilar initial conditions may lead to very similar end-states. Hence, a given nursing unit, which at X point in time looks like another unit in terms of a given variable or relation between variables, cannot be presumed to be similar initially. In this hospital study, for instance, because Unit A and Unit B both had the same adaptiveness scores, one cannot know what "produced" these scores.

Using the perspective of what theory guides the research, one might find another way of looking at the problem. In Dubin's (1969) view, theories of human and social behavior address themselves to two distinct goals of science — prediction and understanding. Although these two goals may not be inconsistent or incompatible, the research strategies used to reach them may differ. Prediction attempts to specify outcomes; understanding essentially means knowledge about interaction of units. Thus, the analytical focus, and methodologies used, will differ. To achieve *understanding*, the analytical focus must be on interaction, and some way to study process must be used. To *predict* accurately, the analytical focus must instead be on outcomes in the form of values of the units, or conditions, or states.

In Goldstein and Farlee's study the goals seem to be both understanding and prediction. But as the specific pieces of the research are being reported, it is not clear which goal is the focus at any particular moment.

There is another important issue in the choice of methodology. In the HIS system study, satisfaction of employees is measured by using the average length of employee stay or, alternatively, employee turnover. Considering again the principle of equifinality, we can readily see that this is a good example of a similar outcome having diverse origins or initial conditions. Employees may leave (similar outcomes) for very diverse reasons (dissimilar initial state), especially within a largely female occupation, such as nursing. "Leaving," for nurses, may be initiated by family pressures generated by children, pregnancy, a husband's being transferred, decision to return to school, and so on, and not by the one initial condition of dissatisfaction with employment.

THE SOCIALIZATION MODEL AND SYSTEMS ANALYSIS

Farlee (1970) analyzed the failure of an innovation — computer-generated medication schedules — by using the concept of socialization as the theoretical model. There are serious questions, however, about the adequacy of the socialization model to analyze a social process. When other theory is used, or when one's starting point is subsystem A rather than B, the outcome of the innovation described might be viewed as "success" rather than "failure." Farlee (1970), using theories from Bredemeier, Parsons, and Bales, synopsizes the socialization model in this way:

In essence, changing behavior or accomplishing the incorporation of new behavior in the task structures of individuals being socialized involves an optimal mix of several elements. Ideally, the socializing agents communicate to the targets of socialization expectations and descriptions of desired behavior, insure that there are adequate opportunities for learning and performing the new behavior and communicate the benefits to be gained. The socializing agents should already be significant reference groups, or manage to become important to the individuals so that their expectations and sanctions are meaningful. The socializing targets should expect to receive rewards from significant others if they perform the desired behavior and should expect sanctions if they do not. Alternate means of achieving the desired rewards should not be accessible, and the costs for alternate behavior and for not performing the required behavior should be high while the costs for performing the desired behavior should be low. When new behavior is incorporated into the task structure of individuals, it is performed routinely and without reference to the sanctions of significant others (pp. 1-2).

In this summary of the socialization process Farlee is, in the system term, viewing one subsystem — possibly a higher level or suprasystem — as the socializing force or agent for another subsystem, here the collectivity of nurses within the total hospital. This immediately raises questions about the direction of communication, the role of feedback between the systems, and adaptive mechanisms. If one takes the socialization view, nonconformity to the desired behavior, when found empirically, can only be classified as "deviance." The trouble with deviance as a concept is that it implies something undesirable, dysfunctional, or to be discarded with haste. It implies also a threat to system survival. According to Buckley's (1967) interpretation of Parsons, deviance presents the social system with problems of control,

since, if tolerated, they will tend to change or disintegrate the system. The system thus "meets" these problems through its "mechanisms of control": mechanisms of socialization and mechanisms of social control thus work hand in hand with mechanisms of defense and adjustment in the personality system to motivate actors to conformity with the given system of expectations, counteract deviance and other strains in the system to bring it back to the given state, and maintain the initial equilibrium (p. 25).

The assumption is that if the system's mechanisms of control do not work, disaster or disintegration will occur. It is *this* assumption which I am challenging. Change, or refusal to do what socialization directs, should not be viewed as leading to disaster. Each subsystem in interaction has an equal right to determine goals and objectives; and, further, only when this negotiation of goals occurs as an ongoing process can survival or adaptation be assured.

Critics have often noted the ambiguous status of "deviance" in Parson's social system. Although in some of his work he describes deviance, tension, and strains as integral parts of a social system, nevertheless, his major theme is the portrayal of the "system" as the dominant, legitimatized, institutionalized structure, or "at least with those characteristic structures that *do not include* patterned strains or structured deviance and disorder" (Buckley, 1967, p. 25). The issue, then, is "whether conformity and order, on the one hand, and deviance and disorder on the other, are to be considered on a par as system characteristics or products" (Buckley, 1967, p. 26). If the answer is yes, then the interpretation of this particular series of events in a medication schedule will change. The sequence of events will be described, along with Farlee's analysis of the data and a possible alternative view.

The HIS "medications application" included four features: (1) requisitions printed in the pharmacy for medications not in stock; (2) daily care plans (medication summary for each patient); (3) medication tickets; and (4) medication schedules. Farlee focuses on the fate of the medication schedules. These schedules are printed out at the nursing stations prior to specified administration times. Confirmation of administration of medications by the nurses was expected "within a reasonable time." The confirmation or exception to the schedule was to be phoned to the information specialist in the central pool, using message line numbers as the reference points. Reminder messages were printed at the nursing stations if this process was not completed; one-half hour remained to confirm the medication before the next list was printed. The nurses were told that each day a shift or midnight list of all unconfirmed medications would be printed out at the nursing office. Nursing supervisors would then be armed with "hard data" to enforce expected performance by the nurses.

When the medications application "went live," several new behaviors were required of the nurses. They had to integrate the confirmation or exception step into their routine, rearrange the output so that it coincided with their actual patient assignments, chart administration of medications in a new way, and adjust local (unit) administration times to coincide with the printed schedules. The socialization model, employed by Farlee in her analysis, would suggest that sanctions would have to operate to induce the nurse to make these changes and to provide punishments for not doing so.

Farlee (1970) describes in detail how this application failed, essentially because of the inadequacy of available sanctions for nonconformity to the system. For instance, nurses soon discovered that they really did not have to confirm the schedules, because medications dropped out of the system, even when not confirmed, as soon as the next schedule printed out. This happened because of limits on the computer's storage capacity. Further, the shift audit was never programmed; therefore, nurse supervisors had no way of systematically administering positive or negative rewards. About a year after this application had been live, the data processing department stopped printing the medication schedule on all but the night shift, and efforts to make it work were abandoned. Farlee summarizes, "it would appear that the innovators created a situation in which most of the essentials for successful socialization were missing" (p. 11).

The application of a systems model, however, would suggest success rather than failure. The outcome of changes in the medication format would be viewed as successful adaptation by interacting subsystems. System theory allows the conceptualization of a "spiraling-forward" phenomenon, where change rather than adherence to a status quo position is the order of the day. The new interrelations among parts of the larger system, the hospital, occur at successive steps, with old or discarded modes being left behind. Change is seen as both inevitable and necessary for system survival, not as deviance.

REFERENCES

Buckley, W. *Sociology and Modern Systems Theory*. Englewood Cliffs, N. J.: Prentice-Hall, 1967.

Campbell, D. T. Common fate, similarity, and other indices of the status of aggregates of persons as social entities. *Behavioral Science*, 1958, *3*, 14-25.

Chin, R., and Benne, K. D. General strategies for effecting changes in human systems. In W. Bennis, K. Benne, and R. Chin (eds.), *The Planning of Change*. (2d. ed.) New York: Holt, Rinehart and Winston, 1969.

Dubin, R. *Theory Building*. New York: Free Press, 1969.

Farlee, C. Failure of an innovation: Computer-generated medication schedules. Unpublished paper, Rutgers University, Center for Urban Social Science Research, New Brunswick, N. J., 1970.

Farlee, C., and Goldstein, B. A role for nurses in implementing computerized hospital information systems. Revised version of paper presented at the Meeting of the Hospital Information System Sharing Group, Kansas City, Missouri, 1970.

Georgopoulous, B. S., and Mann, F. C. *The Community General Hospital*. New York: Macmillan, 1962.

Goldstein, B., and Farlee, C. Nursing unit accommodation to technological change. Unpublished paper, Rutgers University, Center for Urban Social Science Research, New Brunswick, N.J., 1970.

Goldstein, B., and Farlee, C. What's in a name: Evolution of an occupation. Unpublished paper, Rutgers University, Urban Studies Center, New Brunswick, N. J., 1971.

Hage, J., and Aiken, M. *Social Change in Complex Organizations*. New York: Random House, 1970.

CHAPTER 20

Systems Approach to Studying Information Processing and Decision Making in Medical Diagnosis

Steven H. Schwartz
Roger I. Simon

One path to reform of medical curriculum, which many urge is necessary to improve teaching of diagnostic skills, is based on the premise that more precise and efficient educational methods could be devised if the process and requisite skills of medical diagnosis were better understood. This paper describes an explicit model of the information processing procedures that constitute medical diagnosis. The model is guiding our empirical investigations, which are aimed at identifying the important components of diagnostic skill. Its unique features are its specificity and comprehensiveness, as well as its integration with current research in cognitive psychology. These characteristics are of prime importance if one accepts the view that science progresses more rapidly through disconfirmation of specific hypotheses rather than confirmation of vague ones (Popper, 1961).

Research Attempting to Assess Diagnostic Performance or Measure It More Precisely

Prominent in this area has been the work of Rimoldi et al. (1958), Rimoldi (1961, 1964), Cowles (1965), Williamson (1965), McGuire and Babbot (1967), Fleisher (1968), and Heifer (1968). Each has developed a number of

This research was conducted with the cooperation of the Division of Educational Services and Research, Wayne State University, School of Medicine. We are particularly indebted to Dr. Joseph Hess, Division Director.

simulated diagnostic problems and associated sets of questions and answers concerning patient history, physical examination, and laboratory tests. Usually, a medical student (or physician) selects items sequentially from a given set of questions until he reaches a diagnosis. Performance is evaluated in terms of both the final diagnosis and the particular sequence of information a subject selects.

Unfortunately, the validity of generalizing results in such tasks is questionable. A situation in which subjects have all the alternatives presented in front of them (McCarthy, 1966), and are often instructed to achieve a diagnosis "as quickly as possible" (Rimoldi, 1964), bears little resemblance to most actual diagnostic problems. However, these studies have even a more basic limitation in that they offer no model or theoretical framework guiding a finer analysis of the specific components of diagnostic performance. Given this lack, they can contribute little toward a more thorough understanding of the processes being assessed.

Research on Synthesizing Procedures

Most investigators of diagnostic tasks conceptualize them as discriminatory tasks, wherein the subjects discriminate by combining numerous cues in some fashion (Edwards, 1972; Goldberg, 1968; Hoffman, 1960, 1968; Kaplan and Newman, 1966; King, 1967; Ledley, 1972; Ledley and Lusted, 1959; Lusted, 1968; Overall and Williams, 1961; Warner, Toronto, and Veasy, 1964; Woodbury, 1970). Physicians (or a computer program) are typically presented with tasks involving a limited number of cues (signs and symptoms) and required to make diagnoses. Various models have been proposed for explaining the observed cue combining processes. They range from linear combinations (Hoffman, 1960) to conjunctive and disjunctive or other nonlinear models (Einhorn, 1970; Slovic, 1966), lens models (Todd and Hammond, 1965), Bayesian computerized models (Edwards, 1972; Kaplan and Newman, 1966; Lusted, 1968; Meehl, 1954), and signal detection models (Lusted, 1972; Swets, 1972).

Currently researchers within this area disagree considerably over the precise nature of the synthesizing processes. A linear combination of weights has regularly accounted for the major proportion of variance in judgments, but a number of investigators have believed that judgmental processes are more complex and have tried to show that nonlinear models account for a significant portion of the variance in judgments (Einhorn, 1971).

The types of synthesizing models listed in the preceding paragraph may not be relevant to identification of the specific elements or steps in the diagnostic process. Models that describe synthesis of cues to produce a judgment in terms of a mathematical function provide data at a level of abstraction far removed from that of an information-processing description. Linear and nonlinear synthesizing models employed by earlier investigators are useful summary descriptions which account for the judgments produced by the processes mentioned above. But these models imply little about the specific processes utilized.

In contrast, process models, such as the one to be presented in this paper, detail the sequence of activity by which one weighs evidence and integrates it to form a diagnostic conclusion of judgment. Modeling at this level of specific information processes is desirable for a number of reasons. First, it allows the investigation of hypotheses at a level compatible with evidence about basic cognitive processes (memory, attention, pattern recognition, concept formation). Second, it anticipates a wider range of educational interventions based on such theorizing, because of the greater specification of the determinants of diagnostic behavior. Third, the model considers synthesis as only one aspect of medical diagnosis; it also stresses the processes by which relevant information is acquired previous to synthesis. By studying synthesis in isolation, the investigator may not only neglect other vital aspects of diagnosis but may easily get a distorted view, since any interactions of other processes involved with synthesis during the diagnosis will not be manifest.

Research on Information-Gathering and Record-Keeping Procedures

A number of investigators have studied the information-gathering aspects of medical diagnosis (which are often initiated with the goal of automating certain aspects of the physician-patient interaction). The reports of Hershberg (1969), Katz, Gurevitch, Peled, and Danet (1969), Kleinmuntz (1969), Collen (1965), Brodman and Woerkom (1966), Anderson and Day (1968), and Budd, Bleich, Sherman, Reiffen, Strong, and Boyd (1969) all detail examples of this approach. In many of these studies, the automated questions and screening devices have been effective, but again the development of these screening procedures has not been based on any theoretical model. Rather, they are patterned after the thoughts and experiences of one or more experts in certain medical specialties. The better procedures (for example, Weed, 1969) have been modified in the light of experience in use. Some screening procedures work better than others, but it is hard to specify why. In seeking to understand the information-gathering aspects of diagnosis, one must develop theoretically-derived screening procedures to supplement those based on "imitation of experts."

Research Based on Comprehensive Models
of Medical Diagnosis

This research is most relevant to understanding the components of medical diagnosis. Few studies in this area have been based on investigating aspects of specific models.

Elstein, Kagan, and Jason (1970) have employed a variety of "simulation" techniques, including videotaping of physicians working up cases with trained actors as patients and medical students' providing information on small cards and laboratory forms. Although their project is still in progress, they have suggested five mental processes involved in medical diagnosis (Kagan, Elstein, Jason, and Shulman, 1970): (1) automatic scanning of patient; (2) focusing on dissonant elements; (3) selectively retrieving information from memory based on associative processes; (4) determining and ordering of tentative diagnostic hypotheses; and (5) formal information gathering and hypothesis testing. While they have not yet presented a formal detailed model they have reported empirical evidence suggesting that important components of diagnosis are being uncovered. For example, significant differences have been found between medical students and experienced physicians in both number and type of diagnostic hypotheses they offer to a standard set of signs and symptoms (Elstein, Loupe, and Erdmann, 1971).

Wortman (1966), following up previous work on representation and strategy in medical diagnosis, has built a model of diagnosis based on observations of an expert neurologist solving a number of diagnostic problems. He has described rules or strategies by which various diagnostic possibilities were eliminated or confirmed and is preparing a computer program based on these rules (Kleinmuntz, 1969, 1970; Wortman, 1966). As in the studies by Elstein and his colleagues, Wortman's work emphasized information-processing and continuous generation of hypotheses.

Gorry (1970) and Gorry and Barnett (1968) also look at medical diagnosis as a problem-solving activity in which information is stored and processed. They divide medical diagnosis into three major components: the physician's general medical experience; the process whereby he assesses the similarity of a partial pattern of signs and symptoms to those of relevant disease prototypes; and the process by which the clinician selects potentially useful questions, laboratory procedures, and the like to obtain more information on which to base a diagnosis. The components stress the sequential contingent nature of medical diagnosis.

The strategy of research in this case is the development of a model based on observation of diagnostic behavior and introspection by expert physicians to be tested by formalization in a computer program, which is then run on actual data.

Gorry and Barnett (1968), testing the program on problems of diagnosis of heart disease, gave a generally efficient and valid performance. However, to make inferences with confidence concerning specific processes in the model, we would need to study diagnostic behavior of a variety of physicians in different areas of medicine.

The limitations faced by previous studies of medical diagnosis are three-fold: (1) many lack an overall theoretical framework within which to evaluate results, (2) processes studied are often so artificially isolated that there is little hope of understanding their operation in the context of an actual diagnosis, and (3) performance is often considered at a level of abstraction far removed from specifying the processes of information gathering and analysis. The model presented below attempts to meet these criticisms.

A MODEL OF MEDICAL DIAGNOSIS

Medical diagnosis is considered a sequential problem-solving process. Like a host of other diagnostic tasks (for example, psychiatric classification, automotive and electronic trouble shooting, debugging a computer program, aerial photographic reconnaissance), medical diagnosis shares a common concern with the identification of the dynamics that underlie overt signs. (Tasks that classify and predict without recourse to any underlying structure are not considered "diagnostic.")

The study of complex human functions has undergone a major revolution within psychology in the past 15 years. The fact that "higher mental processes" are today being studied by a substantial number of psychologists is important, as is the level of detailed analysis being employed, and it reflects appreciation of the richness of the behavior under scrutiny.

The information-processing approach can be seen in research such as that of Amarel (1970); Bruner, Goodnow, and Austin (1956); Feigenbaum (1963); Gregg and Simon (1967); Hunt (1962); Hunt, Marin, and Stone (1966); Mandler (1967a, 1967b); Miller (1956); Miller, Galanter, and Pribram (1960); Newell, Shaw, and Simon (1958); Norman (1969, 1970); and Reitman (1965). This approach better than any other rubric expresses the spirit of this revolution. It conceives of man as an active selector and encoder, organizer and storer, transformer and manipulator, decoder and retriever of information.

The successes of this approach in leading to a clearer understanding of cognitive behavior, as demonstrated in such tasks as chess (De Groot, 1965); concept-formation (Hunt, 1965; Reitman, 1965); general problem-solving

(Ernst and Newell, 1969; Fikes, 1969; Simon and Newell, 1971; Schwartz and Fattaleh, 1972); and understanding of language (Neisser, 1967) encourage a similar approach to diagnostic behavior.

Our model attempts to describe process at a greater level of specificity than previous works. The following assumptions underlie the model.

First, medical diagnosis is a sequential problem-solving activity. That is, judgments and decisions are occurring all the time. The physician does not passively gather information, stop, synthesize it all, and reach a diagnosis. Rather, the decision processes occur repeatedly.

Second, medical diagnosis, as only part of the total medical problem, is affected by that total context. Therefore, treatment, prognosis, and other such effects must enter into any model of diagnosis. (Crichton, 1970, in a recent popularized account, describes a case where necessary immediate treatment to reduce a very high fever serves to destroy evidence as to its etiology.)

Third, the central construct that guides all phases of the physician's diagnostic behavior is his hypothesis pool, i.e., the current set of diagnostic or problem possibilities. These may vary in specificity from "there may be some disorder in the pulmonary system" to some specific disease.

Fourth, the manner in which a physician represents or structures his medical knowledge will affect almost all aspects of his diagnostic behavior. As such it is a major construct. Differences in structure are expected between various disciplines or specialities in medicine as well as between practitioners.

The model, depicted in Figure 1, can be described as follows:

1. The process is started by the initial complaint (taken broadly to include conditions and setting of patient contact as well as specific complaint).
2. At this point, on the basis of the data from the initial complaint, patient contact, and his structure of knowledge, the physician already begins to formulate hypotheses at some level of specificity.
3. These hypotheses are combined with previous hypotheses in the hypothesis pool. The first time through the model, there is nothing stored in the pool for the case under study.
4. On the basis of this hypothesis pool—the physician's current notions as to the clinical problems of the patient, and his knowledge of medical prognosis, and treatments—he decides whether diagnosis should be halted, at least temporarily, and treatment instituted. If treatment is instituted, the response must be evaluated and a decision made as to whether further diagnosis is necessary. If treatment is not begun, the physician stops. If it is, he records the data involving the patient's response to treatment and goes back to stage 2. If he is not ready to institute treatment, the physician seeks to identify and remember the evidence relevant to hypotheses currently being considered.
5. If there is no more evidence he can conceive of as being relevant in any

way to the hypothesis he has been considering, then the physician searches his knowledge structure first for other diagnostic possibilities at the same level of specificity at which he has been operating. If this is unsuccessful, he goes on to less specific diagnostic possibilities. If he cannot propose any other hypotheses and cannot think of additional evidence relevant to the hypotheses he is considering, then the diagnostic process stops. If, however, at any level he can think of new diagnostic possibilities, he goes back to modify the hypothesis pool and attempts to identify evidence relevant to that new pool. If such evidence can be identified, he goes on to step 6.

6. At this point the physician retrieves from memory his current plan for screening patients. This current plan in early stages of diagnosis may be primarily some set of fixed screening procedures—but, as the process continues, modification in the "standard" screening procedure takes place.

7. The physician then scans the current screening or evidence-gathering plan and any modifications made which seem necessary to the obtaining of relevant evidence.

8. The evidence-gathering plan is then executed, and some questions, observations, or tests are made, if possible, and the data from these observations are stored and used to modify the hypothesis pool, step 2. If relevant evidence is identified but for some reason all the observations, questions, and tests relevant to that evidence cannot be executed at that time (too dangerous, too costly, inability of the patient to communicate), the next question in the evidence-gathering plan is executed and the results stored. If no more questions or tests exist, the physican may search his medical knowledge for any new hypotheses. As before, if one is found, the process continues. Here again at a number of points the physician may consult texts, other experts, etc. for aid. This is understood and not explicitly depicted in the model, for purposes of simplicity.

We view medical diagnosis as a dynamic sequential, iterative problem-solving activity in which the practitioner must elicit and combine diverse information to reach an appropriate conclusion or course of action. The process includes the assessment of initially presented information, such as how much, what type, and how reliable. This initial information interacts in some fashion with the doctor's relevant knowledge to determine a pattern of search or information gathering. Even at this stage the physician will have formulated tentative hypotheses. Search or information-gathering strategies superimposed on the standard techniques for screening eventually lead the doctor to judge that he has sufficient information. Often he must terminate the search when he has all the information he can reasonably expect without taking further measures that would be too costly. (Unless otherwise specified, the term cost or costly is used in its broad sense throughout this paper and

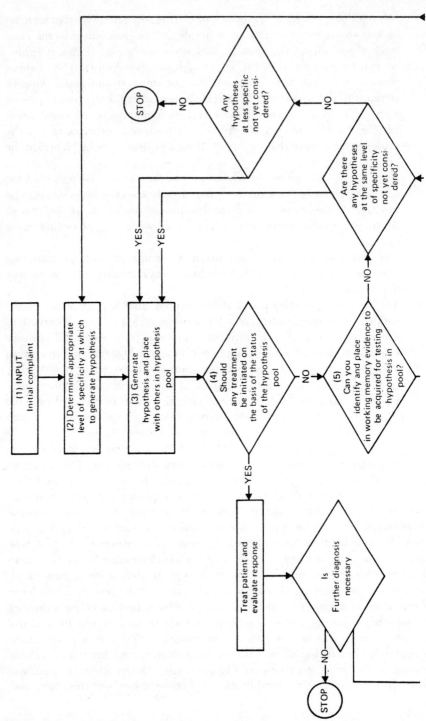

(1) INPUT
Initial complaint

(2) Determine appropriate
level of specificity at which
to generate hypothesis

(3) Generate
hypothesis and place
with others in hypothesis
pool

(4)
Should
any treatment
be initiated on
the basis of the status
of the hypothesis
pool

(5)
Can you
identify and place
in working memory evidence to
be acquired for testing
hypothesis in
pool?

Treat patient and
evaluate response

Is
Further diagnosis
necessary

Are there
any hypotheses
at the same level
of specificity
not yet consi-
dered?

Any
hypotheses
at less specific
not yet consi-
dered?

STOP

STOP

YES

YES

NO

NO

NO

NO

YES

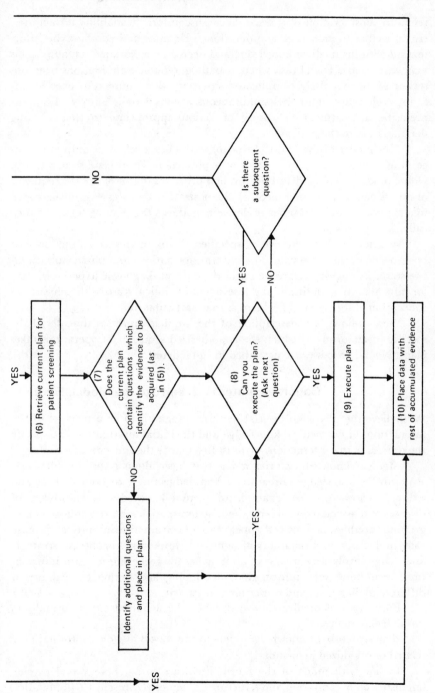

Figure 1. Flow diagram of medical diagnosis model.

YES

NO

(6) Retrieve current plan for patient screening

(7) Does the current plan contain questions which identify the evidence to be acquired (as in (5)).

Identify additional questions and place in plan

NO

YES

(8) Can you execute the plan? (Ask next question)

YES

NO

Is there a subsequent question?

YES

(9) Execute plan

YES

(10) Place data with rest of accumulated evidence

YES

refers to consideration of time, danger, discomfort, availability of facilities, etc. as well as to monetary considerations.) He must now combine this information with his medical knowledge and decide on a working diagnosis. This may vary from a trivial task where a pathognomonic symptom has been observed to an extremely complicated situation where numerous cues of differing reliability must be simultaneously considered. Finally, he must prescribe a treatment or course of action appropriate to that working diagnosis and to the patient.

This model of diagnostic behavior embodies a set of assumptions concerning successive cognitive states of a physician. To give substance to this global model, we must detail the implicit processes that produce the sequence of events culminating in a medical diagnosis. These processes must be stated in a rigorous enough manner to derive hypotheses that can be tested empirically.

We view each of the events specified in the model as a function that operates on various inputs producing transformations in a physician's cognitive state. By cognitive state we mean the current diagnostic hypothesis, plan for information acquisition, and assessment of clinical status—the physician's conception of the medical problem at that particular time.

What follows is a description of the implicit processes that detail the specific mechanisms by which the hypothesized sequence of operations takes place. Not all processes are described in equal detail.

Cognitive States (1, 2, and 3 of the Model, Figure 1)

We believe the physician's initial tentative diagnosis to be a product of the interactions of his medical knowledge and the initial information. To specify the process in this interaction, we must first clarify the concepts involved.

Medical knowledge varies along two basic dimensions: breadth and structure of knowledge. Breadth of knowledge refers to both factual and strategic knowledge. An example of factual knowledge is knowledge of diseases and associated problems, body systems, etiologies, relations to demographic variables, actions and appearance of certain diseases, prognosis, base rates of diseases, and reliability of signs and symptoms. In contrast, strategic knowledge involves awareness of approaches for gathering relevant information, combining such information, developing and testing the validity of differential diagnosis, and prescribing treatment.

Both types of medical knowledge may be acquired either formally or through experience.

The structure of knowledge refers to the way(s) in which information is related or organized in memory.

In line with much of the recent literature in psychology and psycholinguistics on organization and structure of memory (Kintsch, 1970; Mandler,

1967b; Reitman, 1970; Shiffrin, 1970), we conceptualize a structure as a network of elements interconnected by various relationships.

Of particular importance in the model is the structure of knowledge concerning diseases and disease classes. This partition of medical knowledge, we hypothesize, will have a direct functional relation to hypothesis-generation, evidence-gathering, and evaluation. Elements within this structure can be conceptualized as groups of related medical evidence with some joint attributes that discriminate them from other classes of evidence. For example, specific diseases, broad disease classifications, syndromes, prognoses (expected evidence forms), and specified modes of mechanical and biochemical functioning all might exist as distinct elements in a disease structure. Each represents a conceptual label applied to a specified set of medical evidence.

Two major classes of relationships help to organize these elements. The first refers to a "superstructure" or "architecture of patterning" among elements. These relationships order elements into a hierarchy of levels, with the result that elements at different levels vary in status from one to another. Status refers to the degree of differentiation that a given level provides to the total set of elements in the structure. For example, one level may contain five general types of etiologies (a very low level of differentiation among elements), while the level "below" it contains elements specifying diseases associated with various organ systems differentiated within each of the five etiologies.

Particular antecedents of a superstructure are difficult to specify. Presumably primacy effects would be important; that is, to what the student is first exposed, as with a specific textbook and instructor classifications.

The second class of relationships refers to organization at a specific level in a hierarchy. That is, elements at the same level may be found to be exemplars of a particular class of elements, such as diseases related to particular anatomical location, diseases related to specific body systems, diseases that specify particular etiologies. It is important to note that differences in structures can thus be characterized by a variety of orders (vertically in the superstructure) of element classes.

The model asserts that the physician's medical knowledge (particularly his conception of disease classes) interacts with the "initial complaint" to produce a set of initial diagnostic hypotheses at some level of specificity. Although little evidence is available on the specific processes physicians actually employ in this interaction, the following is one possible specification.

The general conception of process we use is one of "structure-guided search." The initial complaint evokes a search process through the disease structure, the goal being the identification of a set of diagnostic hypotheses (region within the structure) to be considered. Such an appropriate level is found when the information available is sufficient to begin discriminating among alternative branches at that level. For example, with a few common symptoms (general shortness of breath, slightly blue skin tone, malaise), the physician quickly notes that many specific diseases may exhibit such signs. Therefore, the symptoms do not discriminate between elements at a level organized around specific diseases. The physician then moves up his super-structure to the next level of organization, for instance, diseases classified as pertaining to specific organs, and again notes that these symptoms char-acterize diseases of a number of organs. He then moves up to the next level, for example, to diseases classified by gross etiologies (metabolic, congenital, carcinomic, traumatic, infectious, etc.), where the information available can discriminate among disease classes so that one or two possible sets of diseases fitting the case can be identified. This, then, is the initial level of hypothesis specification. From here the physician plans to gather additional evidence that will allow him to work down the structure toward more specific diagnostic statements.

As a further example of how two different structures might generate different initial hypotheses, consider the following example. Figures 2 and 3 indicate two possible disease structures physicians can use to organize their knowledge of diseases, as, for instance, when given the following initial set of information:

Patient is a male, 55 years old, who has been a heavy smoker for 20 years. He has been complaining of shortness of breath, has two-pillow orthopnea (needs two pillows to sleep on at night), and appears to have a slight bluish color.

In the first structure (Figure 2), there is not enough information at level three or two to discriminate among elements. However, at the first level the doctor can single out diseases of cardiovascular system as the most likely disease class. The physician would then explore possibilities of diseases associated with particular tissue types (pericardium, myocardium, etc.).

In the second structure (Figure 3), the effective discrimination also begins at the second level where the class of diseases labeled "Acquired, degenerative" becomes the set most likely to contain the final diagnosis.

Since the model is an iterative one, the discrimination procedure occurs again and again, whenever evidence is evaluated and the set of diagnostic hypotheses under consideration is changed. More specifically, once an ap-propriate region of the structure is associated with a set of hypotheses under consideration, the patient's response (or other form of evidence) is evaluated as follows:

If the question (observation, test, etc.) is relevant to the current hypothesis pool, then the evidence is evaluated as to whether it definitely confirms some hypotheses in the pool, definitely rejects hypotheses in the pool, introduces more specific or refined hypotheses into the pool, adds to the likelihood of some hypotheses in the pool, lowers the likelihood of some hypotheses in the pool, or adds rival hypotheses to pool. (This list of considerations may be processed to different extents and in different orders by various physicians, and would have consequences for the quality and completeness of diagnosis.)

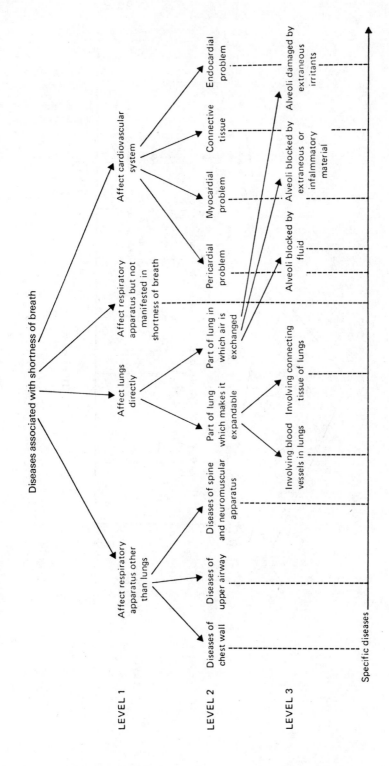

Figure 2. Disease structure A.

237

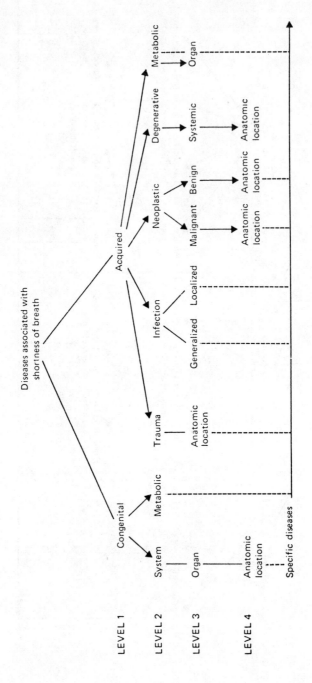

Figure 3. Disease structure B.

238

In contrast, if the question is not relevant to the current hypothesis pool, we hypothesize that the following will occur. A "normal" response will be briefly noted and placed in inactive memory. A response that differs significantly in terms of the implications above may also initiate an additional set of independent hypotheses central to this abnormal finding.

The processes result in the modification of the current hypothesis pool. The modification, in turn, may produce subsequent changes in the plan for information-acquisition as new evidence becomes important.

Criteria for Acting on a Diagnosis (Model State 4)

The model postulates a decision rule that enables a physician to determine whether he has achieved a diagnosis sufficient for treatment. (The model prescribes no special set of processes for arriving at a diagnosis sufficient for treatment but rather the hypotheses pool is continuing to be modified by the processes specified, and reviewed for its sufficiency for clinical interventions.) One of the five following contingencies must be met for this judgment to be made.

1. The hypothesis pool contains a set of one or more diagnostic hypotheses which are confirmed beyond all reasonable doubt. No additional data, given the present state of knowledge in the field, could alter the diagnosis.
2. The hypothesis pool contains a set of one or more diagnostic hypotheses confirmed with high probability. Additional data could alter the probability of the diagnosis, but the expected value of such data is considered too low to warrant further search.
3. The pool contains only hypotheses with functionally identical implications for treatment, and the expected value of additional data for generating additional hypotheses is low.
4. The pool contains only hypotheses in which the treatment indicated by one hypothesis has no adverse implications for the patient if some other hypothesis is correct, and the expected value of additional data for generating additional hypotheses is low.
5. The status of the illness is too serious to risk delay in instituting therapy, even though the working hypothesis is at a very general level — for example, a patient is in shock and some type of treatment is mandatory, even though the physician does not as yet know the etiology of the shock. (Even in emergency situations the physician cannot act unless he has some general hypotheses as to the underlying nature of the problem. The parameters above may change, i.e., the threshold in terms of likelihood of diagnosis may be lowered but the basic processes remain the same.)

If any of these five conditions are met, the doctor prescribes some treatment. If none are met, the physician can seek more information, postpone diagnosis until information requested but not yet available is acquired, wait until the disease dynamics change the state of the patient, call in additional consultants, or treat critical manifest symptoms, despite lack of a precise diagnosis.

A diagnosis sufficient for clinical intervention is not necessarily a final diagnosis. Often the diagnosis at the time of first intervention is modified or confirmed by treatment effects and/or disease prognosis. Opportunities for information-gathering after the first contact with the patient use contingent questions directly related to the current set of hypotheses.

Identification of Evidence Related to Hypothesis Pool (Model State 5)

Once he has identified an appropriate region within the knowledge structure and its diagnostic hypotheses, the physician next seeks to identify evidence needed to narrow down the hypothesis pool to a final diagnosis. The selection strategies employed will probably vary with the physician and the area of medicine.

Considerable literature in psychology deals with types of selection strategies that subjects employ in concept-attainment tasks (Bourne, 1965, 1966; Bruner, Goodnow, and Austin, 1956; Hunt, 1965; Laughlin, 1965, 1966; Schwartz, 1966). Although these tasks differ in many ways from medical diagnosis (the number of relevant dimensions is usually specified), no treatment and cost problems are involved and combinatory rule or type of concept usually is specified. Some of the same basic strategies are employed in both tasks because both require a search of numerous cues and combinations of cues to determine the etiology underlying the behavior. Therefore, one would expect to find strategies employed by physicians differing on such variables as: extent to which he seeks evidence (at various stages) to eliminate alternative hypotheses versus confirming a favored one, extent to which he seeks evidence characteristic of one or a small class of diseases versus many diseases, degree to which he shifts from one physical or biochemical system to another, degree to which he shifts from one problem hypothesis to another (that is, is evidence gathered first on one problem and then another, or on many simultaneously), and level of analysis (at various stages) at which he seeks evidence (what system involved or specific disease type).

States of Model 6-10

Once one has selected evidence relevant to the hypotheses being considered, the actual gathering of information (signs, characteristics, etc.) indicative of that evidence may still be a formidable task. Errors on differences in quality of performance may take place at a numer of points. The physician may elicit information that is not related to the hypothesis he is testing in the way he assumes, may elicit unreliable information, or may require information available only through expensive, dangerous tests. With laboratory data in particular, additional special problems arise such as time delay, cost, risk to patient, and reliability of evaluation of the test. Many of these problems are not unique to laboratory data but appear there more dramatically than in actual physical examination.

Information-gathering in diagnosis is a structured procedure. The formal sequence, which begins with history and is followed by systems review, physical examination, and laboratory tests, is normally followed with minor modifications. In addition to these gross consistencies, there is considerable standardized screening within each of the sources of data. This consideration is often ignored (at the cost of loss of generalization to actual diagnostic practice) by investigators concerned with assessing diagnostic performance (Cowles, 1965; Rimoldi, 1961; Williamson, 1965). A need to achieve a diagnosis as quickly as possible, for example, may lead to simplification in the analysis of thought processes; however, because of the artificial nature of the task, the utility of the procedure is questionable. Each physician develops a specific information-acquisition plan. This consists of a standard screening procedure modified constantly by the need to acquire information about one's present diagnostic hypotheses.

EMPIRICAL RESEARCH

While the specification of the particular strategies physicians employ in each of the above operations has not yet led to empirical research, initial investigation has begun (Schwartz and Simon, 1972). Our proposed information-processing model of medical diagnosis attributes a central role to the physicians' medical knowledge. We consider this representation to be useful in determining almost all aspects of diagnostic behavior. The research reported here is a first attempt to characterize differences among experienced physicians, residents, and senior medical students in their representation of medical knowledge. We derived our approach from those of Miller (1967), Kintsch (1970), and Wortman (1966). In both linguistic structuring and knowledge of neurologic disease, the researchers found the notation of a hierarchical tree to be a useful conceptualization of knowledge.

Wortman's study dealing with an aspect of medical knowledge is particularly significant, even though his results are inconclusive. The three neurologists who served as subjects sorted 115 cards, upon which common neurologic diseases were printed, into major disease categories and subcategories. Analysis of the sortings revealed considerable agreement by the neurologists in major classifications, but little finer differentiations. Whether this resulted from the particular area of medicine and diseases under study, the instructions, or the set of subjects is difficult to conclude from this one study.

Nevertheless, we used the free-sorting procedure and hierarchical-tree notation in our experiment.

Subjects were 13 experienced physicians (all five years or more beyond residency), 13 residents, and 13 senior medical students, all affiliated in some way with Wayne State University College of Medicine. A set of 59 disease names or 69 symptom names associated with shortness of breath (printed on small index cards, prepared in each case by a physician and two residents with the aid of some standard medical text) were displayed before each subject, whose task it was to sort the diseases (or symptoms) into groups which had "diagnostic utility" *to him.* (Subjects were explicitly asked to avoid "textbook classification" unless such divisions actually had such diagnostic usefulness for them). Multiple cards were available for each disease (symptom) so the subjects could use an item more than once. After sorting the cards, the physician was asked if he could further divide these items into finer groupings that still had considerable diagnostic utility. Each was permitted to continue until no further divisions seemed useful to him. He was then asked to characterize each pile of symptoms (diseases), indicate if any groups were more closely related than other groups, and describe his overall sorting scheme if possible. Each subject underwent both the disease and symptom sort, the order of presentation being determined by chance. All subjects sorted cards individually in sessions of up to two hours with two male nonmedical experimenters. The design, thus, was two-factor; one with the type of sort, a within factor, and the three levels of medical experience, a between factor.

In addition to recording card placements and time, we made tape recordings of each subject's responses to brief questions about how he sorted the diseases and symptoms. Hierarchical trees were constructed for each disease, and symptom sorts and further dependent measures were derived from this tree (Table 1).

Physicians took considerably longer than either residents or students to do the sorts ($p<.01$). However, the physicians used more cards, and thus they include the same symptom or disease in more clusters ($p<.05$). This might be expected as a consequence of their greater experience.

Variables 3 and 4 in Table 1 are measures applied to the co-occurrence matrix of each item with every other item over groups of subjects. They reflect the extent to which members within a group agreed in their placement of diseases or symptoms. Although agreement at this fine a level was only moderate, both residents and students agreed more with their colleagues than the experienced physicians did among themselves. Interesting differences in performance between the disease and symptom sorts appeared the first time in the measures derived from the hierarchical trees for each subject (variables 5, 6, and 7). The physicians used more levels of classification on the

Table 1. Mean Scores on Dependent Measures for Physicians, Residents, and Medical Students[a]

Variable	DISEASE SORT				SYMPTOM SORT			
	Physicians	Residents	Students	Total	Physicians	Residents	Students	Total
1. Time in minutes	47.38	23.92	28.53	33.28	49.30	32.23	37.84	39.79
2. Number of items used	71.60	61.15	58.10	63.62	97.80	90.15	80.00	89.32
3. Inter-correlation of co-occurrence matrix with other Ss in peer groups	.28	.36	.40	.09	.09	.31	.25	.05
4. Reduced uncertainty of co-occurrence matrix	.85	1.32	1.34	.97	.55	.87	.70	.54
5. Number of levels in tree	2.69	1.61	2.00	2.10	1.53	1.30	2.00	1.61
6. Number of terminal nodes	13.38	11.85	11.54	12.26	12.38	13.31	11.46	12.38
7. Number of branches per node	6.63	8.29	4.80	6.57	9.38	11.67	5.86	8.97

[a] n = 13 per group

diseases than on symptoms (2.69 to 1.53, $p<.01$), while other groups remained about the same. There were no significant differences between groups on total number of clusters generated, and all ranged between 11 and 14. If one considers differentiation however, in terms of the number of branches from each node (variable 7) in the tree, the results were quite different. The symptom sorts always branched more than the disease sorts ($p<.001$), while in both cases the residents' sorts branched most, physicians next, and students least — all differences significant ($p<.05$). Both the residents and the physicians differentiated diseases and symptoms to a greater extent than did students, with the physicians using more vertical differentiation (deeper trees with more levels) and the residents preferring horizontal developments (more branches for each node).

There were also significant differences between groups and conditions in the specific types of clusters formed. For example, in the symptom clusters the physicians and residents overwhelmingly preferred "etiological" clusters indicative of either a specific cause for the symptoms (heart failure, emphysema) or a "classic etiology" (infectious, carcinomic, congenital, traumatic, etc.). Students, in contrast, often chose "systems" clusters (cardiovascular, neurologic, pulmonary, etc.). In disease clusters, however, students were much more prone to stick to a classic etiologic scheme, the one usually found in textbooks.

The techniques employed in this study produced many intriguing findings relating to how one type of medical knowledge is represented. However, these findings must be interpreted with extreme caution. Our sample was small and undoubtedly biased, because we could not *force* experienced physicians to serve as subjects and it was limited to only one content area (shortness of breath). Nonetheless, many findings appear worthy of careful further investigation, for example, the difference in vertical versus horizontal differentiation between residents and experienced physicians should have implications for the nature of the diagnostic problem-solving activity each engages in.

This study has shown that knowledge structures can be isolated and classified and that they vary considerably with experience. We are now exploring the mechanisms that produce an interaction between structure and diagnostic process; specifically, we are attempting to relate classes of disease structures to specific evidence-gathering and hypothesis-modification procedures used by physicians. Our goal is to converge on an information-processing model characteristic of superior diagnostic performance. The identification of such processes and their interactions should be a major step toward fundamental reform of curriculum medical education, as well as a considerable advance in the understanding of complex cognitive functioning.

REFERENCES

Amarel, S. On the representation of problems and goal-directed procedures for computers. In M. D. Mesarovic and R. Banerji (eds.), *Theoretical Approaches to Nonnumerical Problem Solving*. New York: Springer-Verlag, 1970

Anderson, J., and Day, J. L. New self-administered medical questionary. *British Medical Journal*, 1968, *4*, 636-638.

Bourne, L. E., Jr. Hypotheses and hypothesis shifts in classification learning. *Journal of General Psychology*, 1965, *72*, 251-261.

——. *Human Conceptual Behavior*. Boston: Allyn and Bacon, 1966.

Brodman, K., and Woerkom, A. J. Computer-aided diagnostic screening for 100 common diseases. *Journal of the American Medical Association*, 1966, *197*, 901-905.

Bruner, J. S., Goodnow, J. J., and Austin, G. A. *A Study of Thinking*. New York: Wiley, 1956.

Budd, M., Bleich, H., Sherman, H., Reiffen, B., Strong, R., and Boyd, G. Survey of Automated Medical History Acquisitions and Administering Devices—Part 1. (HSRD-70-13). Springfield, Va.: Clearinghouse for Federal, Scientific and Technical Information. 1969.

Collen, M. F. Multiphasic screening as a diagnostic method in preventive medicine. *Methods of Information in Medicine*, 1965, *4*, 71-74.

Cowles, J. T. A critical-comments approach to the rating of medical students. *Journal of Medical Education*, 1965, *40*, 188-198.

Crichton, M. *Five Patients: The Hospital Explained*. New York: Knopf, 1970.

De Groot, A. D. *Thought and Choice in Chess*. New York: Basic Books, 1965.

Edwards, W. N equals 1: Diagnosis in unique cases. Chapter 8 in J. A. Jacquez (ed.), *Computer Diagnosis and Diagnostic Methods*. Springfield, Ill.: Charles C Thomas, 1972.

Einhorn, H. J. The use of nonlinear, noncompensatory models in decision making. *Psychological Bulletin*, 1970, *73*, 221-230.

——. Use of nonlinear, noncompensatory models as a function of task and amount of information. *Organizational Behavior and Human Performance*, 1971, *6*, 1-27.

Elstein, A. S., Kagan, N., and Jason, H. Methods for the study of medical inquiry. Paper presented at the annual American Psychological Association Convention, Miami, 1970.

Elstein, A., Loupe, M., and Erdmann, O. An experimental study of medical diagnostic thinking. *Journal of Structural Learning*, 1971, *2*, 45-53.

Ernst, G., and Newell, A. *A Case Study in Generality and Problem Solving*. New York: Academic Press, 1969.

Feigenbaum, E. A. Verbal learning and concept formation. In E. A. Feigenbaum and J. Feldman (eds.), *Computers and Thought*. New York: McGraw-Hill, 1963.

Fikes, R. E. REF-ARF: A System of Solving Problems Stated as Procedures. Unpublished AIG technical report, Stanford Research Institute, 1969.

Fleisher, D. S. Composition of small learning groups in medical education. *Journal of Medical Education,* 1968, *43,* 349-355.

Goldberg, L. R. Simple models or simple processes. *American Psychologist,* 1968, *23,* 483-496.

Gorry, G. A. Modelling the diagnostic process. *Journal of Medical Education,* 1970, *45,* 293-302.

Gorry, G. A., and Barnett, G. O. Experience with a model of sequential diagnosis. *Computers and Biomedical Research,* 1968, *1,* 490-507.

Gregg, L. W., and Simon, H. A. Proves, models and stochastic theories of simple concept formation. *Journal of Mathematical Psychology,* 1967, *4,* 246-276.

Heifer, R. E. Assaying the process of reaching a clinical diagnosis. Paper presented at the 7th Conference on Research in Medical Education, Houston, November, 1968.

Hershberg, P. L. Medical diagnosis: The role of a brief, open-ended medical history questionnaire. *Journal of Medical Education,* 1969, *44,*293-297.

Hoffman, P. J. The paramorphic representation of clinical judgment. *Psychological Bulletin,* 1960, *57,* 116-131.

———. Cue-consistency and configurality in human judgment. In B. Kleinmuntz (ed.), *Formal Representation in Human Judgment,* New York: Wiley, 1968.

Hunt, E. B. *Concept Learning: An Information Processing Problem.* New York: Wiley, 1962.

Hunt, E. Selection and reception conditions in grammar and concept learning. *Journal of Verbal Learning and Verbal Behavior,* 1965, *4,* 211-215.

Hunt, E. B., Marin, J., and Stone, P. *Experiments in Induction.* New York: Academic Press, 1966.

Kagan, N., Elstein, A. S., Jason, H., and Shulman, L. S. A theory of medical inquiry. Paper presented at the annual American Psychological Association Convention, Miami, 1970.

Kaplan, R. J., and Newman, J. R. Studies in probabilistic information processing. *IEEE Transactions on Human Factors in Electronics,* 1966, *HFE-7,* 49-63.

Katz, E., Gurevitch, M., Peled, T., and Danet, B. Doctor-patient exchanges: A diagnostic approach to organizations and professions. *Human Relations,* 1969, *22,* 309-324.

King, L. S. What is a diagnosis? *Journal of the American Medical Association,* 1967, *202,* 714-717.

Kintsch, W. Models for free-recall and recognition. In D. Norman (ed.), *Models of Human Memory.* New York: Academic Press, 1970.

Kleinmuntz, B. *Clinical Information Processing by Computer.* New York: Holt, Rinehart, and Winston, 1969.

———. Clinical information processing by computer. In *New Directions in Psychology IV.* New York: Holt, Rinehart, and Winston, 1970.

Laughlin, P. R. Selection strategies in concept attainment as a function of number of persons and stimulus display. *Journal of Experimental Psychology,* 1965, *70,* 323-327.

———. Selection strategies in concept attainment as a function of number of relevant problem attributes. *Journal of Experimental Psychology,* 1966, *71,* 773-776.

Ledley, R. Syntax directed concept analysis in the reasoning foundations of medical diagnosis. Chapter 9 in J. A. Jacquez (ed.), *Computer Diagnosis and Diagnostic Methods.* Springfield, Ill.: Charles C Thomas, 1972.

Ledley, R. S., and Lusted, L. B. Reasoning foundations of medical dagnosis. *Science,* 1959, *130*, 9-21.

Lusted, L. B. *Introduction to Medical Decision Making.* Springfield Ill.: Charles C Thomas, 1968.

————. Observer error signal detectability and medical decision making. Chapter 3¡ in J. A. Jacquez (ed.), *Computer Diagnosis and Diagnostic Methods.* Springfield, Ill.: Charles C Thomas, 1972.

Mandler, G. Organization and memory. In K. W. Spence and J. Spence (eds.), *The Psychology of Learning and Motivation.* Vol. 1. New York: Academic Press, 1967. (a)

————. Verbal learning. In *New Directions in Psychology III.* New York: Holt, Rinehart and Winston, 1967. (b)

McCarthy, W. An assessment of the influence of cueing items in objective examinations. *Journal of Medical Education,* 1966, *41*, 263-266.

McGuire, C., and Babbot, D. Simulation technique in the measurement of problem solving skills. *Journal of Educational Measurement,* 1967, *4*, 1-10.

Meehl, P. E. *Clinical Versus Statistical Prediction.* Minneapolis: University of Minnesota Press, 1954.

Miller, G. A. Psycholinguistic approaches to the study of communication. In D. L. Arm (ed.), *Journeys in Science.* Albuquerque: University of New Mexico Press, 1967.

————. The magic number seven, plus or minus two: Some limits on our capacity for processing information. *Psychological Review,* 1956, *63*, 81-96.

Miller, G. A., Galanter, E., and Pribram, K. *Plans and the Structure of Behavior.* New York: Holt, 1960.

Neisser, U. *Cognitive Psychology.* New York: Appleton-Century-Crofts, 1967.

Newell, A., Shaw, J. C., and Simon, H. A. Elements of a theory of human problem solving. *Psychological Review,* 1958, *65*, 151-166.

Norman, D. A. *Memory and Attention: An Introduction to Human Information Processing.* New York: Wiley, 1969.

Norman, D. Models of human memory. In D. Norman (ed.), *Models of Human Memory.* New York: Academic Press, 1970.

Overall, J. E., and Williams, C. M. Models for medical diagnosis. *Behavioral Science,* 1961, *6*, 134-141.

Popper, K. R. *The Logic of Scientific Discovery.* New York: Harper & Row, 1961.

Reitman, W. R. *Cognition and Thought.* New York: Wiley, 1965.

Reitman, W. What does it take to remember? In D. Norman (ed.), *Models of Human Memory.* New York: Academic Press, 1970.

Rimoldi, J. J. D., Devane, J. R., and Crib, T. F. Testing skills in medical diagnosis psychometric laboratory. Unpublished paper, Loyola University, 1958.

Rimoldi, H. J. A. The test of diagnostic skills. *Journal of Medical Education,* 1961, *36*, 73-79.

————. Testing and the analysis of diagnostic skills. In J. A. Jacquez (ed.). *The Diagnostic Process: Proceedings of a Conference Sponsored by the Biomedical Data*

Processing Training Program of the University of Michigan. Ann Arbor: University of Michigan Medical School, 1964.

Schwartz, S. Trial-by-trial analysis of processes in simple and disjunctive concept-attainment tasks. *Journal of Experimental Psychology,* 1966, *72,* 456-465.

Schwartz, S. H., and Fattaleh, D. L. Representation in deductive problem solving: The matrix. *Journal of Experimental Psychology,* 1972, *95,* 343-348.

Schwartz, S. H., and Simon, R. I. Differences in the organization of medical knowledge among physicians, residents, students. *Journal of Structural Learning,* 1972, *3,* 23-36.

Shiffrin, R. M. Memory search. In D. Norman (ed.), *Models of Human Memory.* New York: Academic Press, 1970.

Simon, H., and Newell, A. Human problem solving: The state of the theory in 1970. *American Psychologist,* 1971, *26,* 145-160.

Slovic, P. Cue consistency and cue utilization judgment. *American Journal of Psychology,* 1966, *79,* 427-434.

Swets, J. Signal detection in medical diagnosis. Chapter 2 in J. A. Jacquez (ed.), *Computer Diagnosis and Diagnostic Methods.* Springfield, Ill.: Charles C Thomas, 1972.

Todd, F. S., and Hammond, K. R. Differential feedback in two multiple cue probability learning tasks. *Behavioral Science,* 1965, *10,* 429-435.

Warner, H. R., Toronto, A. F., and Veasy, L. G. Experience with Baye's theorem for computer diagnosis of congenital heart disease. *New York Academy of Science Annals,* 1964, *115,* 558-567.

Weed, L. L. *Medical Records, Medical Education and Patient Care.* Cleveland: Case Western Reserve University Press, 1969.

Williamson, J. Assessing clinical judgment. *Journal of Medical Education,* 1965, *40,* 180-187.

Woodbury, M. Applicability of Baye's theorem to medical diagnosis. Paper presented at the conference on The Diagnostic Process, Ann Arbor, June, 1970.

Wortman, P. M. Representation and strategy in diagnostic problem solving. *Human Factors,* 1966, *8,* 48-53.

CHAPTER 21

Systems Approach in Studying Medical-Surgical Nursing

A. Dawn Zagornik

My interest in the systems approach to learning began with a presentation by a neurophysiologist who applied the cybernetic-control system to the regulation of blood pressure as a method to learn and understand the physiology of blood flow and pressure. In the four years since, my interests in the application of "systems approach" to the teaching-learning process has grown and I have introduced this broad field to graduate students in medical-surgical nursing.

Leonard C. Silvern (1964) has defined a system as the "structure or organization of an orderly whole, clearly showing the interrelations of the parts to each other and to the whole itself" (p. 1). The systems approach is the application of analysis and synthesis to a system. According to Silvern, it is the exhilarating process of analysis-synthesis, which he terms anasynthesis, that offers the real meaning of the systems approach.

Because analysis and synthesis are two major cognitive skills which the graduate student is striving to augment, the systems approach seems then to be a highly effective learning technique.

In using a systems approach to the study of medical-surgical nursing over a period of several years, we have developed a five-phase plan as one way of introducing and applying the systems approach (Figure 1).

The first phase is the black-box concept. The black-box concept usually is not well known to graduate nursing students, but it is basic to the understanding of any type of systems approach and contains the basic components of any system, i.e., input, regulation-control, and output. The black-

Phase 1. Black-Box Concept

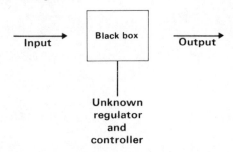

Phase 2.   Relationship Between Black-Box Concept and Nursing Process

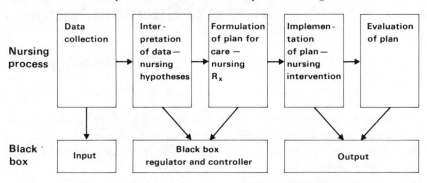

Phase 3.   Relationship Between Nursing Process and Control Systems

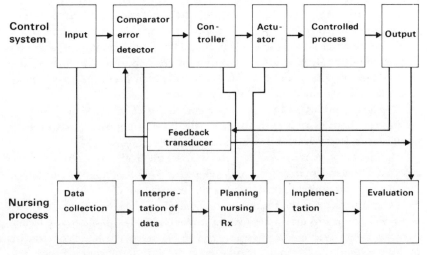

Figure 1.   Application of systems approach to the study of nursing.

Phase 4. Modifications of Concepts and Development of "Own" Models

Phase 5. Apply Systems Approach to the Study of a Particular Nursing Problem
Use to explore:

Biological State	*a.*	Patients' anatomical and physiological deficits or alterations imposed by stressor.
Psychosocial State	*b.*	Patients' psychosocial response to illness state and relationship to biological state.
Physical State	*c.*	Patients' setting within health care system. Setting analysis to determine assets and deficits within system and how it affects the patient and nursing health care.
Provision of Care	*d.*	Interaction of health professionals and paraprofessionals in providing care.

    1. Assessment of patient
    2. Plan of care
    3. Implementation of care plan
    4. Evaluation of care

    e. Relationship of nursing prescription to other health professionals' prescriptions of care.

Figure 1 *(continued).*

box, which stimulates the learner to investigate the possibilities of what is regulating and controlling the system, is the unknown at the time of inquiry, no matter what the boundary of the system might encompass.

Phase 2 relates the black-box concept to the nursing process, and it highlights the relationship of regulation and control to interpretation of patient data and the formulation of the plan for nursing care. Nursing process implies a systems approach with a built-in feedback mechanism through the validation of assessment and the results of patient care (Figure 1).

I view the nursing process as a system of related processes and as a model for the understanding and solution of problems (Berggren and Zagornik, 1968). The understanding and application of the nursing process is the beginning of a systems approach to nursing.

Phase 3 probes the relationship between the nursing process and the control system. The control system offers another conceptual approach which, when related to the nursing process, strengthens the students' frame of reference in studying variables that impinge on or augment the provision of nursing care. Using the control systems model in relation to the nursing process provides the learner with another set of components—i.e., error detector, controller, actuator, controlled process, and feedback transducer —which provides more concrete links to the study of the chain of events in giving patient care.

When the student has progressed to this point in his understanding, during Phase 4 he can then explore other systems models—with the notion that the student will then be stimulated to use his own selected system throughout Phase 5.

Phase 5 is the actual application of the systems approach to the study of a particular nursing problem or situation. This can include exploration of the patient's biological state, psychosocial state, physical state, and the provision of care. The systems approach is particularly useful in determining nursing prescriptions and how the prescriptions relate to those determined by other health professionals.

The reaction of students to the five-phase plan has been both favorable and unfavorable. Students who support the use of the systems approach as a learning tactic in the study of nursing have commented that the systems approach has: "made me think and feel more responsible for learning; provided a framework from which to detect gaps in analyzing a situation—check and balance system; made me feel creative in studying about nursing; forced me to make interrelationships between categories of phenomena; helped me identify gaps in knowledge and set new learning objectives; helped me develop a more holistic approach to patient care; helped in developing writing skills—forces you to become more concise and critical."

Comments from students who do not favor this approach include: "made me work so hard in developing models, etc., that I lost interest in major nursing problems; tormented me; took too much time in developing models which could have been spent on gaining more factual knowledge; became too dehumanizing to use in studying about people; challenged my creativity, and I block every time I try to even think about systems and model development; provided me an intellectual game which doesn't help to solve problems of the real world."

I believe there are several advantages to the systems approach as seen from the chair of the faculty. Systems approach helps the student further his ability in (1) abstract and critical thinking; (2) conceptualizing patient situations and the nurses' role; (3) problem solving; (4) developing conceptual frameworks on which to hang observed phenomena and to organize patient data; (5) verbal and written communication; and (6) asking pertinent questions.

I believe that several questions concerning the systems approach should inspire further research:

First, what type of person(student) is more likely to benefit from systems approach method in developing cognitive abilities for analyzing problems of nursing and the health delivery system?

Second, what are some alternative teaching strategies to help students utilize the systems approach, including model development?

Third, would the systems approach provide a framework from which to build coordinated curricula for the health professional schools?

Fourth, what evaluative devices can be used to measure the effect of the systems approach method on the development of cognitive abilities?

Fifth, what joint research efforts between schools could be developed to utilize fully the exciting, creative work of graduate students who are working with and through systems concepts?

Clinical faculty in medical-surgical nursing are extremely enthusiastic about using a systems approach to the analysis of clinical nursing problems. This approach seems to provide more adequately than others for inclusion of the many complex variables that contribute to the state of the patient, and to help the student identify not only which variables need to be manipulated to alter the patient's state, but also the ways in which those variables might be manipulated.

However, faculty share with students the conviction that the benefits of the systems approach to teaching medical-surgical nursing must be measured in terms of whether or not it ultimately contributes to improved nursing care of those patients whose problems are the concern of the graduate student. That judgment can best be rendered after systematic investigation of the process and the product of applying the systems approach to the analysis of real nursing problems.

REFERENCES

Berggren, H. J., and Zagornik, A. D. Teaching nursing process to beginning students. *Nursing Outlook,* 1968, *16*(7), 32-35.

Silvern, L. C. A general system model of public education K-12. *Teaching Aid News,* 1964, *4* (18), 1-20.

CHAPTER 22

Systems Approach to Measuring and Modeling the Patient Care Operation

Richard C. Jelinek

In the most general sense, performance measures for any operational system are based on the quantity and quality of output for a given quantity and quality of input. Thus, to develop an understanding of performance of an operational system, it is necessary to have some understanding of the system as a whole; that is, not only must one be able to quantify output, but one must also be able to quantify inputs and understand their relationship to output.

Our interest lies in measuring perormance of the patient care service, and to begin we must conceptualize this operational function. This function will be restricted to include hospital activities that have direct bearing on the patients' welfare yet are not performed by, nor are the direct responsibility of, the physician. The activities of this function may be grouped into two major categories: patient care activities that fulfill the specific and implied orders of the physicians, and patient care activities that are executed at the discretion of the nursing staff. The patient care system is illustrated in Figure 1.

Figure 1 describes pictorially the elements that make up the patient care function and their relationships to each other. The most important feature of the model is a transformation of inputs (resources) into units of output or service. This transformation may be affected by other factors such as work load, organization, and environment. A somewhat more specific description (Jelinek, 1967) of the elements that make up the patient care system follows:

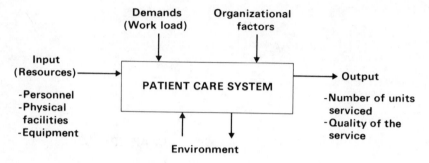

Figure 1.   Patient care system.

1.   Input factors
Describing the resources used in the patient care operation. Resources include personnel and the physical facilities.
2.   Organizational factors
Describing the form of organization used in the patient care operation. Factors in this category include the rules and policies used, the degree of "work specialization," and the type of supervision.
3.   Work load factors
Determining the work load that any group of patients imposes on the resources, i.e., on the personnel and the physical facilities. These factors are characterized by the number and condition of patients to be serviced.
4.   Environmental factors
Representing elements other than those that are a part of the patient care operation, but which have effects on, or are affected by, patient care. Factors in this category include hospital organization, medical staff organization, and other hospital departments or activities.
5.   Output factors
Describing the outcome of the patient care system in terms of both quantity and quality of the patient care rendered.

PERFORMANCE MEASUREMENTS

Once we have the conceptual model for the patient care function, we can look at the problem of measuring the performance of the system. Although measuring performance is multidimensional in that no single, universally acceptable measure exists, there are two basic dimensions to the performance measurement of a patient care system. The first is measurement of the "quantity" of output for a given amount of resources used. In the health field this measure has generally been reversed by measuring the resources used to produce some unit of output or service. We will use the latter approach in our discussion and will refer to this class of measures as the "quantity" measures. The second is measurement of the quality of the service rendered. This measurement consists of two basic factors: the waiting time for the service, and the standard of performance of the service. We shall refer to these measures as the "quality" measures.

The "quantity" measure relates the input (resources) used to some number of units of an output. (Note that this is the concept of efficiency; however, whereas efficiency is defined as the ratio of output to input, we define "quantity" as the ratio of input to output.) This measurement may be in terms of (1) man-hours per unit of output; (2) materials per unit of output; (3) facilities used per unit of output; or (4) dollars per unit of output. The Hospital Administrative Services of the American Hospital Association data on nursing service operations are examples of measurements presently used to represent this dimension of performance in the area of patient care function. These include direct cost of nursing service per patient-day, nursing man-hours per patient-day, and the nursing service man-hours per bed per day. Note that the direct cost of nursing service, nursing man-hours, and nursing service man-hours are considered as measures of input, whereas patient-days and beds per day are considered as measures of output.

The "quality" measure concerns itself entirely with the output. Measurements in this area are directed toward quantifying the "goodness" of the output. In the area of patient care this may represent the degree to which needs for care are identified; for example, recognizing that a surgical dressing needs changing; or the waiting time for the service, i.e., the delay from the time the need to change the dressing has been recognized to the actual changing of the dressing (waiting times could be negative [for instance, a drug administered before it was due]); or the quality associated with the accomplishment of a particular task (how carefully sterile techniques are practiced in the changing of the surgical dressing).

The relationship between the quantity measure and the quality measure is of utmost importance. It can be argued that in general the relationship between quantity (resources per unit of output) and quality (the goodness of the output) takes a form similar to that illustrated in Figure 2, assuming that resources used are optimally utilized with respect to the quality measure. This relationship is such that quality increases as the quantity increases; however, it increases at a disproportionate rate in that added increments of quantity yield progressively fewer increments of quality. The nature of the relationship can be argued theoretically; it is also, at least in part, supported empirically.

Arguing theoretically, we can consider a group of patients requiring various needs for care. It can be assumed that the various needs take on different levels of importance as far as the patients' welfare or the quality of the care is concerned. We can assume that a unit of a resource, say a staff nurse, is introduced to satisfy the needs. For this nurse to maximize her effectiveness

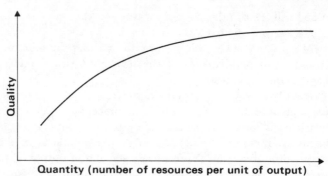

Figure 2.   Quantity versus quality relationship.
(From R. C. Jelinek, F. Munson, and R. L. Smith, *Service Unit Management: An Organizational Approach to Improved Patient Care* [Battle Creek, Mich.: W. K. Kellogg Foundation, 1971])

in terms of quality, she would allocate her time to meeting those needs which in a given time period contribute the most to quality. If an additional nurse (a unit of the same resource) were added, then the needs that she can satisfy, on top of those already satisfied by the first nurse, will contribute to improving quality; however, the contribution will be less than that made by the first nurse (since the first nurse already met those needs that contribute most to quality). In a real situation, it can be assumed that the nurses' training and experience are aimed at the development of the ability to recognize needs and identify those needs that have greatest value to the patient and thus to quality. With this ability to identify importance of needs (although probably quite imperfect) nurses will allocate their time to patients' needs in a way that will maximize quality. If, in fact, this argument holds in a real situation, the relationship between quantity and quality will follow the form illustrated in Figure 2.

It has been shown empirically that as the size of the nursing staff (resources) is increased, the amount of time that the staff devotes to patient-centered activities follows a relationship similar to that described above. Added increments of nursing staff yield progressively fewer increments in the amount of time devoted to patient-centered activities (Jelinek, 1967). Although in patient-centered activities time is not a measure of quality, it probably does have some relationship to it.

Another empirical study shows the existence of this form of a relationship between staff size and quality of care, when quality is based on an index derived from sample observations measuring the presence or absence of certain attributes associated with quality of patient care (Jelinek, Munson, and Smith, 1971).

It has also been suggested that too large a quantity of a resource may result in inefficiency because of congestion and may actually result in a reduction in quality as quantity is increased beyond a certain point. In answer to this suggestion, it could be argued that optimal utilization of resources has not been attained and that it could be attained by "storing" the additional resources, by not using them. In any event, no matter which argument is accepted, its effect will have little influence on our formulation.

The line in Figure 2, which relates quality to quantity, represents the case in which the resource in question is optimally utilized with respect to quality. Thus, in any actual operational situation we would generally not expect to have an optimal utilization of resources; and, consequently, an operational point would be expected to fall below the optimal relationship line, at some point, as illustrated in Figure 3. This point indicates that a hospital with a quantity level of R_1 attains a quality level of Q_1. For this organization it could be expected that by changing the quantity level, quality would follow some curve similar to that labeled actual relationship. Thus, according to Figure 3, one possible managerial strategy — illustrated by arrow A — would be that directed toward increasing quality by simply increasing the quantity of resources while keeping utilization at the level of the existing system. An alternative strategy would take the course represented by the arrow B. This course better utilizes existing resources to improve the level of quality. (Note that this strategy results in a more efficient system without, however, any cost savings in terms of resources.) Another strategy could take the course of arrow C. This is a strategy directed toward better utilization of resources so as to keep the quality at its original level Q_1, while at the same time reducing the quantity of resources from R_1 to R_2. Another possible

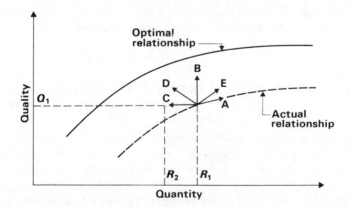

Figure 3. Operational interpretation of the quantity versus quality relationship. (From R. C. Jelinek, F. Munson, and R. L. Smith, *Service Unit Management: An Organizational Approach to Improved Patient Care* [Battle Creek, Mich.: W. K. Kellogg Foundation, 1971])

strategy is one following course D, where utilization of resources is improved and resources are reduced, although not to the same degree as they were in strategy C. This results in a reduction of cost as well as an increase in quality (the most desirable situation to hospital administrators). A strategy following the arrow E is also feasible.

Although, in practice, there are no universally acceptable measures for the quality of the output for the patient care operation, a number of approaches exist that are useful for obtaining estimates for it. These approaches fall into two categories: those using expert judgment, which although subjective in nature have been found and have a certain degree of consistency (Georgopoulos and Mann, 1962); and those using a quality index based on sample observations directed toward measuring the presence or absence of certain attributes associated with quality of patient care (Veterans Administration, 1966; University of Michigan, 1967). In addition to these two existing approaches, there are numerous research efforts directed toward the development of measurements for patient care (Nadler, Huber, and Sahney, 1967).

The quantity to quality relationship is one that must be considered in any systematic analysis of the patient care operation. Quality must be taken into account whenever an understanding of the effect of any change in the operational patient care system (inputs) on performance of the system (outputs) is needed. For certain interests in the area of patient care systems, the problems of measuring quality can be reduced by establishing certain quality standards and simply making sure that as quantity is changed the quality standards are met. Another approach, frequently used when focusing on changes in the inputs to the system, is to predict their effect on the quantity measures associated with output by arguing that quality measures will be at least as good after the change as they were before—an argument that may be logically sound, although in general quite subjective in nature.

PERFORMANCE MEASURES IN THE MANAGEMENT CONTROL OF THE PATIENT CARE SERVICE

With at least a conceptual understanding of the patient care system as an input-output model able to predict the output for any given set of inputs (elements influencing the output), we can now use this information in the management of the patient care system. The management control system is illustrated in Figure 4. The major element of this control system is the patient care system itself, whose output is monitored to determine the level of performance. This measurement is then compared with a preset standard. On the basis of this comparison, the inputs to the patient care system may be adjusted to correct for any discrepancy.

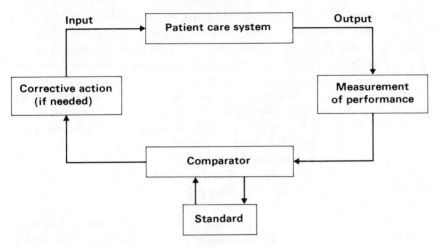

Figure 4. Patient care management control system.

The first requirement for the patient care management control system is the ability to measure the system's performance in terms of both quantity measurements (resources per unit of output) and quality measurements (waiting time for service and standards of performance) while also taking into consideration the relationship existing between them. The lack of presently available and satisfactory quality measurements greatly hinders the development of a truly objective management control system. Nevertheless, a great deal can be done with existing measures. Furthermore, the research developments in this area are progressing rapidly as new measurement techniques are becoming available.

The second requirement for the patient care management control system is that of establishing the standards associated with each of the performance measures. In the patient care system described previously, a multitude of inputs (resources in terms of personnel, facilities, and equipment; type of patient; organizational factors; and the type of environment) determine the output. Furthermore, the system's performance can be measured in terms of quantity as well as quality measures. These characteristics of the system result in the need for a variable standard, one that can adjust itself to the appropriate level of the input factors. As an illustration, a standard stated in terms of the number of hours of the professional nursing staff's time per patient-day is certainly affected by the type of patient being cared for. Thus, such a standard needs to be adjusted for the type of patient being given care.

The problem, then, becomes one of determining the influence of the relevant factors on the performance measure in question. For example, a standard for the professional nursing hours per patient-day may be influenced by each of the following: number of nonprofessional nursing personnel; mix of nonprofessional nursing personnel; design of the patient unit; degree of support from other departments; number of patients; degree of illness of patients; and type of equipment and use of facilities. Furthermore, the problem of determining the influence of the above factors is made more complex by the fact that actual performance, in addition to the above factors, could be influenced by factors such as style of management, utilization of personnel (scheduling and assignment of activities), and productivity of personnel. These are factors that probably should not be determinants of a standard; however, their influence on performance may have to be understood before an understanding of the overall system is attained. Although some research has already been done in the area of determining the influence of organizational factors on performance (Jelinek, 1964), considerably more work is needed.

The type of nursing statistics provided by hospital administrative services (HAS) are examples of the performance measurement portion of the management control system. These statistics provide, for example, quantity measures on a total hospital basis. HAS also provides an indication of a standard, which is based on the actual performance of other hospitals in a similar size category. This standard, however, is not corrected for the effect of the many factors influencing performance that have been mentioned above, nor does it correct for the level of quality.

To illustrate the use of performance measurement in the management control of the patient care service, we can use an application dealing with control over the utilization of nursing personnel.

Management Control in the Utilization of Nursing Personnel

Scheduling of hospital personnel, especially nursing personnel, is generally considered by hospital administrators as one of the most important problems in the operational management of hospitals. Numerous studies (Connor, 1960) verify the lack of coordination and control in this area. To illustrate the problem, let us look at the relationship between work load (reflecting the need for nursing care) and staff size in a community general-care hospital. Figure 5 illustrates this relationship.

Similar relationships have been observed in a number of other hospitals. This relationship is fairly typical. Each point on the graph represents the work load and staff for one patient unit over a single day shift (week days only). If some control directed toward adjusting staff to meet the work load were in operation, the points on the plot would fall along a line with a positive slope (represented by the dashed line in Figure 5). As can be seen from the distribution of points, no such relationship exists. If anything, a relationship

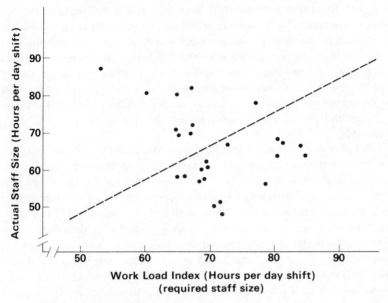

Figure 5. Work load versus staff size.

associated with an entirely undesirable situation—increasing the staff as work load decreases—is observed.

The first concern regarding the acceptance of the relationship between work load and staff size is the question of how work load is measured. Although no universally accepted measure of work load exists, some rational measures are available. One of the simplest is based on the number of patients on a patient unit. The commonly used indicator of nursing hours per patient-day is based on this measure. This may be useful if a particular unit or group of units contains a fairly homogeneous group of patients (patients with relatively similar demands for nursing care). A somewhat more sophisticated measure is based on the classification of patients into categories reflecting the need for care. A work load index is derived by a simple formula utilizing the number of patients falling into each of the categories. This approach was first developed in 1960 by Connor and has since been refined by others. Figure 5 uses this as a measure of work load. There the appropriate weights given to each class of patients which is needed to derive the index were derived for the hospital used in the illustration. A very similar graph results if the number of patients is the only indicator of work load. Also, although the number of hours of total nursing personnel (including R.N.s, L.P.N.s, and Nurses' Aides) is used in Figure 5 as a measurement of staff size, a very similar result appears if the staff size for each kind of personnel is plotted against work load separately.

The relationship between work load and staff size illustrates the lack of management control in one aspect of the patient care operation. However, for this particular problem the basic requirements for a management control system are available to hospitals. In the first place, measurements exist for both inputs—the work load and staff size—and for output or performance —the degree to which work load and staff are matched. Other factors that may be considered important in relation to this problem, such as the degree to which an individual nurse can remain with the same patient, are also measurable.

In the second place, a number of procedures to control this relationship have been proposed and, in certain instances, implemented. These include (1) selective placement of admissions—the information used in assigning a patient to a particular hospital bed can include information relating to existing work loads and nursing staff on each patient unit—and (2) centralized "float" nursing staff—a staff of nurses can be "floated" in the hospital in such a way as to optimize the matching of available staff to work load. It is anticipated that in general the "float" staff needs to represent less than 10 percent of the total nursing staff in order to control major discrepancies between staff size and work load. I propose that such a system should use a nurse preference profile for deciding on assignments; this profile would describe a nurse's preferences for floating or permanent assignment, as well as how and where she would prefer to float.

This operating scheduling system would utilize the concept of the management control system described earlier. The standard would consist of an acceptable limit on the discrepancy between staff size and work load.

REFERENCES

Connor, R. J. A hospital inpatient classification system. Unpublished doctoral dissertation, Johns Hopkins University. 1960.

Georgopoulos, B. S., and Mann, F. C. *The community general hospital.* New York: Macmillan, 1962.

Jelinek, R. C. *Nursing: the development of an activity model.* (Doctoral dissertation, University of Michigan.) Ann Arbor, Mich.: University Microfilms, 1964, No. 65-5912.

———. A structural model for the patient care operation. *Health Services Research,* 1967, *2,* 226-242.

———, Munson, F., and Smith, R. L. *Service Unit Management: An Organizational Approach to Improved Patient Care.* Battle Creek, Mich.: W. K. Kellogg Foundation, 1971.

Nadler, G., Huber, G., and Sahney, V. A study concerning the measurement of quality of patient care. Paper presented at the 18th Annual Conference of the American Institute of Industrial Engineering, Toronto, May, 1967.

University of Michigan quality sampling instrument. Unpublished instrument, University of Michigan, Bureau of Hospital Administration, 1967.

Veterans Administration. Nursing care quality control plan. Unpublished paper, Department of Medicine and Surgery, 1966.

PART SIX

Systems Concepts
to Systems Utilization

The preceding sections have been devoted to discourse on general systems theory, discussions on the health care subsystems, and presentations dealing with current research activities in health care systems. This section focuses on the problems and concepts involved in systems analysis, systems synthesis, and systems evaluation. In broad terms, the papers concern the implementation of systems design and performance evaluation.

The first paper deals with several of the research activities of the Applied Health Services Research Center, Indian Health Service, United States Department of Health, Education, and Welfare. After pointing out several of the constraints and problems of existing health care delivery systems, Berg goes on to state that the mission of the Center is to develop, test, refine, and demonstrate optimal ways of planning, budgeting, implementing, and evaluating service programs for almost 450,000 American Indians and Alaskan natives. The overall research program is described as being functionally stratified into three classes of activity: program operations, program development, and community development.

Program operations focus on projects designed to improve the efficiency of operations within the present structure of the system. Thus, current resources and configurations are accepted as givens or constraints and attempts are made within these constraints to improve the system. The author describes the development of an outpatient computer simulation and a project in hospital nurse scheduling as examples of this strategy.

Program development is concerned with the generation and demonstration of optimum mechanisms of planning, budgeting, implementing, and evaluating specific health programs. In this approach, current system configurations are not givens or constraints. The development of a health information center is used as an example of this approach. This system is designed to contain medical, educational, social, and environmental data which are to be retrieved and summarized as needed to provide individual health care. The stated goals for such a system include improvement of medical care, relieving health care professionals of data preparation activities, development of episodic care programs, and determining information needs. The author also describes the medical record, the development of systems criteria, the problems of field observation of health care data, and the interdepartmental communication of such data. Berg concludes his discussion of program development with an example of the implementation of the information in the field of public health nursing and several other problematic situations.

The community development program is described by Berg as having as its goal the development of mechanisms or the provision of training to provide American Indians and Alaskan natives with the necessary skills or techniques for roles to be played in the health field. This activity attempts to develop human and natural resources at the local community level. Berg concludes his paper with a summary of several years' research in this activity.

Klir begins his paper by defining " 'synthesis' as a general name referring to every procedure by which some elements are put together in a certain way so as to form a whole with certain specific properties." The author then discusses the concepts of systems variables, variable resolution levels, external versus internal variables, and system behavior.

According to Klir, the problem of system synthesis, given a particular system behavior, is to determine an organization by which the behavior can be implemented. This organization consists of a program and a structure of the system, the former being variable and the latter being constant. Therefore, it becomes clear that there is no unique solution to appropriate systems analysis. However, as one places more constraints on the system, fewer solutions are possible.

The author suggests that the problem of system synthesis requires two steps. First, a well-defined formulation of the specific problem of synthesis should be formed. Second, the problem should be solved by formal procedure.

Klir concludes his paper with an illustration of the problems and procedures of systems synthesis as applied to a health care system.

The final paper in this section concentrates on concepts and problems in the test and evaluation of health systems. According to De Greene, " 'Test' or 'testing' usually refers to the real-time collection of data under a plan and under controlled conditions. 'Evaluation' then refers to the a posteriori interpretation of the test results." After describing the problems of implementing system evaluation, the author suggests that "test and evaluation are integral parts of the development of new systems, the modification of old systems, the juxtaposing of systems, and the determination of the effects of new operational plans, programs, and strategies."

In discussing methodology, De Greene points out that the criterion problem is a key issue in systems testing and evaluation. It is made clear that operational criteria may be quite elusive regardless of whether we refer to performance, effectiveness, benefits, or costs. This seems to be true because of the multidimensional nature of the phenomena with which health systems must deal.

In addition to presenting various operations research techniques which are applicable to systems evaluation, the author outlines several specific approaches to the evaluation of health systems. Among the approaches discussed are classical experimentation, system comparison with analysis, cybernetic sampling and modeling, and simulation in the development of a health care delivery system.

De Greene concludes with a presentation of the reasons for the lack of impact of systems knowledge on the health field. One reason cited for this deficiency is the focus on aggregating health information and a consequent de-emphasis of the microaspects of health care such as diagnosis, effective treatment, and human engineering of facilities. According to De Greene, other problems stem from historic, legal, bureaucratic, and cultural factors. Finally, part of the problem lies in the complexity of the health system, which makes systems prediction difficult when using single measures of performance.

CHAPTER 23

Applied Systems Analysis in the Delivery of Comprehensive Health Services

Lawrence E. Berg

The applications of the techniques of systems analysis and operations research should aid in the development of a systematic approach to improving the effectiveness of the delivery of health services. Such an approach was used by the Indian Health Service of Tucson, Arizona, of which I am a staff member.

At the outset this service shared the limitations common to most health care delivery systems. These include:

1. Traditionally established agencies and professional disciplines will not surrender jealously guarded prerogatives and areas of responsibility in the presence of overwhelming evidence of duplication or gaps of services.
2. Individuals tend to view themselves as specialists and to accept personal responsibility only for their narrow field of interest.
3. Many community agencies seem to have the intent of providing coordinating functions and, with few exceptions, do not commit themselves to the provision of direct service.
4. Many communities lack a common set of objectives for the development of all members of the community.
5. Professionals are insecure about the role of the consumer in establishing policy for the delivery of comprehensive services and fear that consumers will override professional judgment.
6. Professionals fear "over-the-shoulder" monitoring of the quality and quantity of professional services.
7. Practitioners seem unwilling to develop measurements that evaluate the

impact of services on the health of the recipient.

8. Health care personnel seem willing to accept or work around mediocre performance.

These problems are generally being recognized as insurmountable obstacles to the development of a single truly comprehensive health delivery system for many communities. These same problems, to varying degrees, have operated in the system of delivering comprehensive health services to both American Indians and Alaskan natives.

The Indian Health Service was organized in 1955. The initial priorities were overcoming critical staff shortages and improving facilities. By the early 1960s, substantial gains had been made. The planners recognized, however, that staff and facility resources alone would never be fully adequate to meet Indians' health needs if the delivery of health services continued to be based on the traditional patterns. This awareness brought expanded efforts to improve the organization and management of health resources. The staff knew that the effectiveness of health programs is directly related to the amount and type of involvement on the part of the health services' consumers — in this instance, the Indian people.

A training center and the Health Program Systems Center were established in Tucson in July 1967 for the purposes of increasing health resource effectiveness and Indian involvement in health planning. To further accelerate achievement of these goals, the Office of Research and Development was established in July 1969. This office brought together within one organizational structure the Health Program Systems Center, the Desert Willow Training Center, and the Papago Health Program, which serves as a demonstration model for the systems being developed.

The mission of the Office of Research and Development is to develop, test, refine, and demonstrate the optimal ways of planning, budgeting, implementing, and evaluating the comprehensive health services program for individual and community health services for more than 450,000 American Indians and Alaskan natives. It also seeks to make available to the Indians and Alaskan natives training programs designed to provide skills in their management of health services and specific occupational roles in the health field.

A component of the office is a demonstration health delivery program which provides services to the approximately 7,000 Papago Indians in south central Arizona. The health delivery system is a component of the Office of Research and Development on the organizational chart. But, in practice, the system has a direct delivery staff of about 100 workers operating a health system out of three community health centers and a 50-bed hospital, supported by a 24-member research staff of systems engineers, systems analysts, mathematicians, statisticians, information scientists, behavioral scientists, a research physician, a pharmacist, and a nurse. Health service research needs and their respective priorities are being identified through the actual process of planning, delivering, and evaluating health services in the demonstration program.

We have functionally stratified our approach into three classes of activities: program operations, program development, and community development. Program operations include projects designed to improve the efficiency of the present structure of the Indian Health Service. Characteristically, these are short-term efforts designed for immediate implementation.

Program development focuses primarily on the development of optimum mechanisms of planning, budgeting, implementing, and evaluating specific health programs in areas such as tuberculosis, health of infants, or specific mechanisms required of a health care delivery system, such as information processing.

Community development includes the means for providing the Indians and Alaskan natives with specific skills needed for the management of health services or performance of a specific role in the health field.

In the program operations sector we are accepting the current resources and their existing structure as a given, and are attempting to immediately improve the efficiency of that system. This approach has two obvious advantages:

1. It is dealing with health managers "where they are at," with problems that they currently feel are of high priority.
2. It provides short-term, highly visible payoffs that not only benefit the system but also serve to reinforce the presence of the systems analyst on the health team, since he is more than earning his keep for significant contributions.

For example, in the Papago Health Program as in most others, the patients (in 1969) complained about their waiting time in the health centers. In a typical visit of one hour and 54 minutes the patient was being served an average of 19 minutes, and the balance was waiting time to see the nurse, doctor, pharmacist, or x-ray or laboratory staff. Since 48 minutes of the average total of one hour and 35 minutes of waiting time was spent waiting to see the nurse, the clinic staff asked what reductions in total waiting time would be possible if an additional nurse was added to the staff? A second question was what would be the impact of setting up an appointment system for the patients? Others were interested in the potential impact of adding additional examining rooms without changing staffing. A computer simulation of the outpatient department was built to answer these questions. By December 1969 the model had been given a preliminary test, and it was deemed ready for its first application under field conditions. The model was designated for use at an outpatient clinic on the San Xavier Indian Reservation near Tucson.

In four weeks, data on some 400 patients were collected. By use of automatic time printers located throughout the clinic, patient movements were followed, patient waiting times were obtained for each clinic service point, and consumption of staff time for each patient service was calculated. These data were introduced into the computer program which "operates" the model. A government report on the project emphasized that the outpatient clinic simulation model gave a good approximation of actual clinic conditions in terms of staff and patient waiting times and service times, and in terms of the flow of patients through various clinic stations.

The results of the study were as follows:

1.   The model confirmed prior impressions about waiting times and service times. Total times for services provided at the clinic were generally about 5 to 30 minutes, with an expected mean of 18 minutes. Waiting times were from 5 minutes to two hours, with an expected mean of one hour and 35 minutes total per clinic visit.

2.   The clinic served less than 25 patients a day, and each member of the professional staff performed more than one function, i.e., the physician was also the pharmacist, the nurse was also the laboratory technician. Because of this, and because one action frequently had to await completion of a prior action by the same person, one service that required an exceptionally long time caused long back-ups for many following patients. The model reflected this situation, and it was sensitive to the infrequent but serious impact of one patient's wait upon other patients' waiting time.

3.   Probably the most significant finding was that most patients arrived about the same time, early in the clinic period. Until this heavy backlog of patients was served, excessive waits occurred for most patients. Scheduling appeared to be one possible solution (*Simulation of Outpatient Flow,* 1970).

Of ten hypothetical questions initially asked of the simulation about reducing the patients' waiting time, two changes promised benefits in modifying the operations of the outpatient clinic. The presence of an additional nurse would reduce patient waiting time from one hour and 35 minutes to 52 minutes. Even more significantly, if 25 percent of patients' visits could be scheduled, average waiting time for all patients, the simulator predicted, would be reduced by more than 30 minutes, even without adding more staff. In reality, we found that only 16 percent of the patients adhered to a schedule, which nevertheless reduced average waiting time for all patients by 30 minutes and also greatly reduced the overtime the clinic staff worked. Ten clinics have been surveyed to date, with ten more scheduled in the future. Attendance at these clinics, which are located throughout the western United States and Alaska, ranges from 35 to more than 300 patients a day. This project illustrated how one can systematically analyze several facets of a health care delivery system with a tool sufficiently flexible to apply to various kinds of outpatient clinics being operated in the Indian Health Service. As an example of the speed with which this tool can be used, the computer can simulate thousands of staff-patient encounters or outpatient visits to test the proposed hypothesis about the predicted impact of the patients' waiting time resulting from a proposed change in staffing or structure in less than a minute in computer time, costing approximately ten dollars.

Another project in the program operation sector was the naive entry of one systems engineer into the field of scheduling of hospital nurses. In working with the nurses in the Papago Hospital, the engineer found 29 "rules" that had to be followed for scheduling fewer than 29 nurses to provide 24-hour, seven-day-a-week coverage. There has to be a registered nurse on every shift and a Papago-speaking person on each shift. Mrs. A. could only work days because of a transportation problem; Mrs. B had no telephone and, therefore, was not available for callback, for example. A straightforward assignment problem seemed an easy chore for a good systems man. His first attempt was to determine the optimal solution for the scheduling problem. After several tries, he discovered that there was no solution of any kind to the problem, and this gave administrators a fresh insight into the administrative problems of nursing supervisors. Based on the responses of the individual nurses, we assigned "disutility weights" to all possible types of assignments. We then constructed an assignment mechanism that equalized the disutility points among all the nursing staff and minimized disutility for the total nursing unit, when compared to former operations. Since this tool provides an objective basis for recording and evaluating the subjective feelings each nurse has about the various types of assignments, the senior nursing staff believes that it has potential for use by a local director of nursing in developing staffing patterns. But perhaps more significantly, the current tool, or a modification thereof, has a high potential for use by multi-state area supervisors of nurses for investigating "trouble spots" where morale seems to be suffering.

Potential dollar savings have been identified in several other program operations projects; for example, prepackaged convenience foods in hospital food service, maintenance program mechanisms for water wells, and dental auxiliaries with expanded functions under the supervision of a dental surgeon to replace dentists in the clinic.

In the area of program development, the configuration of the current system is not accepted as a given or constraint. Our major effort in program development has evolved from its initial design as a technical service in program operations.

The computerized information system was initially designed as a central file for medical, educational, sociological, and environmental data about each individual who received medical services or was eligible to receive such services. While the structure of the information system was oriented to the individual, the system would be capable of retrieving and summarizing data by families and/or communities. Teletype terminals at the three health centers would provide access to the data base for the appropriate patient, family, or community information.

This system was a prototype, and it emphasized simplicity of design. It had a modular structure to facilitate future modification and additions to the system, as operating experience dictates. The program attempted to obtain early implementation of the system to test feasibility and acceptability of a central data base, to develop and test concepts and methodologies for system operation, and to start the buildup of a more complete data base of medical and management information.

Some of the reasons why we initiated development of a health information system were:

1. It was generally assumed that the quality of medical care would be improved by a faster flow of information between treatment facilities where a given patient may seek services and health team workers who provide the services.
2. Some assumed that the system would mean less time required of health workers in preparing charts, reports, etc.
3. A few naive ones probably assumed there would be a cost savings through "automation."

We decided not to automate existing manual systems, because workers believed that the existing systems were not responsive to their information needs. By the end of the second year we learned that, for the most part, practicing health workers do not know their exact information needs or cannot articulate them. Most felt they needed only that information which was required, under existing regulations, for reporting to higher-level administrators. We really were not overly concerned about this last class of "needs," but we decided to provide it routinely to satisfy the participating staff.

It became evident that if we wanted to develop an information system that would support a systematic approach to planning the improvement of primary care as well as making each health team member a part of a planned preventive health care program, the programs would have to be developed before the information needs could be identified.

Having a problem-oriented medical record was recognized as a first step in making the limited patient-clinic encounter more efficient. But, we soon learned that providing an automated, problem-oriented medical record alone, without further development of full support machinery, would not provide the assurance that the necessary inputs or follow-up would occur. For example, a current study had identified 42 attempted suicides documented or known among the various agencies on the Papago Reservation for a one year period. Of these 42, eight were officially recorded in the health system.

Health problems, particularly, lend themselves to involvement with community groups other than health professions, such as social service agencies and teachers. These individuals have a role to play in various types of interventions and treatments, but they also have a great deal of information that is helpful in planning a program oriented toward prevention. The observations of a child by a teacher are often the first indicators of serious problems that may be developing. The family visitor, for whatever reason, may very well be the first one to see a situation that will develop into a problem. Similarly, the social welfare worker in the community agency sees many early indications of problems either in their particular clients or other family members.

These observations occur independently of each other and in themselves are often not sufficient to warrant intervention. For example, a lone observation made by a minister may lack the urgency necessary for a referral for help. However, if that observation were backed by observations by a public health nurse, social worker, teacher, or physician, it might be obvious that the patient needed help. Even more confusing is the situation in which several medical specialists are treating the same patient without knowledge of the others' activities. In this situation, any of the physicians might notice some sign of illness that by itself would be relatively unimportant but in conjunction with the observation of another health team member would become serious enough to warrant intervention. This lack of communication makes truly comprehensive medical care impossible. Even more serious is the fact that it precludes the possibility of using many early treatments well within present capabilities.

The present system of medical care has as its only tool for communication the medical chart and independent records maintained by the various field workers. The use or nonuse of these documents is very much left up to each individual. A centralized, problem-oriented medical record is one attempt to change this situation. Each individual health observer, with the use of a problem list, can be aware of the multiple problems that a patient faces. However, this in itself does not guarantee communication among the health team members. A measure of individual initiative still is required.

Working toward the goal of developing a planning process to reinforce early detection and prevention of illness, as well as primary medical care, we needed a mechanism to facilitate communications among different groups and within specific groups. The computer offers the capabilities of memory, coordination, and notification. However, the success or failure of any system in medical care ultimately depends on the cooperation of the individuals involved in its use. We found that the most successful single strategy for making a system work is to involve those responsible for the implementation of the system in the tremendous effort required for its development, to the extent possible.

We have developed so far two key elements of the information system — its problem orientation (placement of medical problems in a larger health context) and a surveillance data base (programmed preventive intervention). The health status surveillance program utilizes a set of minimum requirements for surveillance, including immunizations, skin tests, laboratory tests, X-rays, and special examinations and histories that are scheduled for all members of the service population. Surveillance data requirements have been defined in a manner to aid in early detection of health problems and to identify patients who are at high risk for specific diseases.

The information system gives the physicians and other health team members, as part of the medical summary generated at each patient contact, a current listing of all surveillance procedures that are past due or that are scheduled within the next 12 months. This means that the encounter with a patient may be used to collect and review basic information on health status.

The system also provides periodic reports for public health nurses which list all patients who are delinquent in special surveillance procedures. The public health nurse also periodically receives a list of all recent hospital and outpatient encounters for persons in her district and weekly lists arranged by the community she is to serve that week. Persons are listed because the nurse had indicated at some time in the past the necessity for following up this patient at this point in time; the physician referred her to this person to perform specific tasks; or the patient is overdue for a high priority procedure.

The experiences of developing the public health nursing component of the information system is typical (Brown, Mason, and Kaczmarski, 1971). Of the 1,090 registered nurses in the Indian Health Service, 56 are school nurses, 146 are public health nurses, and the balance work in hospitals or outpatient facilities. When we asked the public health nurses for their assistance in developing that component of the information system, they outlined several problems:

1. Public health nursing data overlap in areas such as clinic operations, school health programs, and disease registers.
2. No standardized terminology exists to describe public health nursing functions, problems, or results.
3. Clinicians show an almost total lack of commitment to preventive health programs.
4. The presence of only a few preventive health objectives, which few other than the field health staff care about, leads to referrals which are primarily extensions of curative treatment plans.

Guidelines for statistical reporting by public health nurses prepared by the National League for Nursing were reviewed, and the procedures used by 26 public health nursing agencies were surveyed. Parts of the systems used in three agencies seemed to have some potential application in the system being constructed. The development of the information system for public health nursing and the corresponding computer programs for processing these data required approximately a year. The first field application of the new data collection tools was primarily designed to test computer programs and processing procedures. After that study, a second field test attempted to determine the problems the field staff would have in recording the services being provided. Because mark-sensitive forms that were optically read were used, many of the early "rejects" resulted from markings that were too light or not completely erased. Other technical problems included items left blank and use of invalidated codes. After nine months of operation, the error rate dropped from an initial rate of 10.9 percent and leveled off at 3.3 percent.

This field test, which included more than 26,000 public health nursing hours, approximately equivalent to 17 years of professional service, revealed surprisingly wide variations in the use of time throughout the testing area. For example, the amount of time spent with individual patients providing direct service varied from 8.1 percent of total time for one field health nursing team to 30.8 percent of total time for a second team. The time spent in office activities or professional meetings varied from one-third of total time for one staff member to more than one-half for another. To the question of whether a person with less training could have conducted part or all of the activities being reported, public health nurses indicated that on the average 48.1 percent of the tasks that they performed during the study period could have been handled, at least in part, by someone with less training, most often naming the licensed practical nurse. The licensed practical nurse indicated that 16.6 percent of her tasks could have been performed by someone with less training, most often the health aide. Some of the functional tools that came out of the pilot project were the development of standardized terminology to describe many of the public health nurse's functions; more specific delineation of the public health nurse's activities, eliminating some of the apparent "program conflicts" with other members of the health team; a clearer perspective of the relative roles of the public health nurses, licensed practical nurses, and field health aides; and further recognition of the need for preventive health objectives and standards.

The automated problem-list program now provides health workers who request it with a consolidated multifacility, multidisciplinary problem list for each patient. Problems included in these lists may be medical, emotional, social, environmental, or educational in nature. A complete and current problem list is included in the medical summary provided to the health practitioner at the point of patient contact. The public health nurse has the option to take into the field either a portable terminal so that she can directly tap the computer data base anywhere there is a telephone or obtain printed and current medical summaries for all the members of families she plans to visit during that field trip. The problem list, which is viewed by peers, forces health team members to view each patient in the context of the total spectrum of his health related problems, rather than in the context of a single complaint or a current problem.

Three additional areas of development will significantly add to the value of the information system. The first area is that of high-risk programs. The surveillance system will be used to identify, in terms of objective criteria, individuals who are at high risk for specific diseases and problems. Staging will be defined for each of the selected disease and problem categories, and minimum standards of care to all high-risk patients will be established.

The second area involves giving the full range of information, as is now provided to physicians and public health nurses, to all members of the health care team. These disciplines, which now receive some information, include mental health workers, alcoholism workers, environmental specialists, health educators, medical social workers, and community health medics.

The third major area involves the bolstering of problem orientation of all members of the health care delivery system. The problem list now used is only the first step in this program. Subsequent steps involve the linkage of treatments and services to specific problems at all encounters. The program also has the ability to store additional classes of information, such as multidisciplinary treatment goals, treatment plans, and progress notes for each problem. With the capability of storing problem-specific treatment goals, plans, and progress notes, the system immediately gains the controversial potential of being able to evaluate these against standards that have been introduced into the logic of the system by the appropriate specialist. The objective of such a function can be easily misinterpreted, but it is not the policing of health disciplines. Rather, it is professional education. If at any time a given professional feels that a treatment plan must be developed which is markedly different from one chosen by the system as introduced by a specialist, he must have the opportunity to follow his own professional judgment and must pull the computer's plug when he feels it is necessary.

In a second major developmental effort, working with the systems engineering department of the University of Arizona, we are developing specific models of trachoma, tuberculosis, and diabetes to assist the Indian Health Service program managers in beginning management programs for optimal use of resources in their area. These models are based in methodology on either mathematical control theory or computer simulation. They consist in each case of a population model that describes the dynamics of the disease process in the population, plus a cost model that reflects social, preventive, and therapy costs. These models are designed to assist the program manager in planning resource allocation for maximum impact.

The trachoma model has been validated on the San Xavier district of the Papago Reservation by four total population screens of this district by an ophthalmologic team. The tuberculosis model has been validated on data taken from recent history of tuberculosis in the United States and data available from the Utah State Health Department. This model is currently being used in the planning of a long-range tuberculosis control program for the Sells Service Unit of Indian Health Service.

The third major class of activities of the office are those involving community development. We recognize that the allocation of health resources must be within the context of total community goals. We also believe that the final authority for establishing these objectives and making the decisions regarding the allocation of resources among competing demands and needs belongs to tribal leadership-management training programs and systems consultation as requested on a wide range of programs, including development of human resources, natural resources, and tribal structure.

We are applying the systems approach to health services delivery with three objectives: to make the existing system more efficient in the short run and, at the same time, to develop mechanisms that will produce a more effective system, and finally to place health needs in the context of total community needs so that appropriate decisions on allocation of resources can be made by those who will be most affected by the consequences of these decisions.

One factor that emerges repeatedly is that the transformation of mechanisms and processes that currently exist into a truly comprehensive health system cannot be easily attained and cannot be approached from any one narrow point of view. The planning and control functions, which information systems serve, require involvement. This involvement cannot be esoteric, academic-like exercises in defining long-term goals that participants discover have no practical implications for health care; rather, it must be concerned with the identification of problems and the resolution processes. Our approach to system development is three-way. In program operations, we are identifying and resolving the problems professionals see as blocks to doing their jobs better. In program development efforts, we are developing, with the clinical specialist, new patterns of health maintenance and intervention to overcome the problems they have identified. In community development, we are repeating the process of developing systems. In working with tribal leaders, we are identifying structural systems that often underlie the kinds of problems now occurring. We feel that with this concurrent three-way approach and our use of traditional tools, as well as systems analysis, we have the potential for building a truly comprehensive health care system.

REFERENCES

Brown, V. B., Mason, W. B., and Kaczmarski, M. A computerized health information system. *Nursing Outlook,* 1971, *19,* 158–161.

Simulation of Outpatient Flow. Tucson: U.S. Department of Health, Education, and Welfare, Health Program Systems Center, 1970.

CHAPTER 24

Systems Synthesis

George J. Klir

"Synthesis" is a general name referring to every procedure by which some elements are put together in a certain way so as to form a whole with certain specific properties. Depending on the nature of the elements involved, synthesis procedures may differ considerably one from another. For instance, the formation of a complex chemical compound by combining several simpler compounds is certainly different from putting together simple elements of thought to make a complex theory. Similarly, the procedure of designing a complex connection of transistors, resistors, capacitors, and other electrical elements to make a television set is, no doubt, different from putting together English words to make a meaningful English sentence or putting together meaningful English sentences to make a meaningful article devoted to, say, systems synthesis.

Although synthesis procedures applied in individual disciplines of science, engineering, or the arts may significantly differ one from another, some general principles have been recognized which are valid for all of them. The study of these general principles, which are referred to as the principles of systems synthesis, belongs to a discipline called the methodology of general systems.

General principles of systems synthesis are elaborated within a particular conceptual framework. Whenever the framework changes, the methodological principles change. I will introduce several concepts that are used in my approach to general systems theory and are described in detail in Klir (1969).

When talking about systems, we can hardly avoid talking about various quantities each of which may assume different values at different time intervals. We shall call them variable quantities or, simply, variables. As examples of variables associated with health care, we can consider various health status indices, the total number of patients in a hospital or in a certain category (emergency, preoperation diagnostics, surgery, intensive or routine postoperative care, convalescence), temperature, blood pressure, and other physiological data of individual patients (Fanshel and Bush, 1970).

Each variable under consideration is measured with certain accuracy and recorded only at certain time intervals. This feature of a variable is called its resolution level. More specifically, the resolution level of a variable is a collection of values (or classes of values) which we distinguish for the variable, together with a set of time instances (or time intervals) at which we measure or observe the variable and record its values. For instance, the number of emergency cases requiring surgery in a hospital is recorded every day with complete accuracy.

Given an object of our interest, like a health care aggregate (I intentionally do not use the term "health care system" in this context), we can define a system on the object by selecting certain variables associated with the object and specifying a resolution level for each of them. The system is thus a partial representation of the object, corresponding to a particular point of view from which the object is observed.

Variables of the system are usually classified as external variables and internal variables. External variables are those which participate in the interaction between the object and its environment; all other variables are internal. We can further classify external variables as input variables and output variables. The output variables are those whose values are determined by the values of the input variables and some properties of the object, but the input variables have values determined by the environment of the object. A record of values of the external variables of a system within a certain period of time represents the activity of the system of that period.

The dependence of the output variables on the input variables, which is either deterministic or statistical, is called the behavior of the system, and the collection of all properties of the object which participate in the creation of this dependence is the organization of the system.

Now, we are able to define very roughly the problem of the synthesis of a system as follows: Given a behavior (variables with resolution levels and dependencies among them), determine an organization by which the behavior can be implemented. To define the problem more specifically, we have to clarify the concept of the organization first.

The organization consists of a variable portion, called the program of the system, and a constant portion, called the structure of the system. The program of the system, which is of no interest in the problem of synthesis, is at any time the present state of all the variables involved (both external and internal), all transitions from this state, the set of next states, the transitions from the next states, and so on.

Systems may contain two kinds of structures. The first kind is the structure of the program of the system, or the structure of states and transitions. This type of structure is defined as the set of all states of the system, together with all possible transitions between these states. I shall refer to this kind of structure as the ST-structure ("state-transition structure").

The second kind of structure is based on the assumption that the behavior of the system is produced as a composition of some simpler behaviors associated with some lower level systems, which are usually called elements of the system under consideration. Each element has its input variables, its output variables, its behavior, and some couplings with other elements of the same system, the environment, or other systems. A coupling between two elements occurs if some output variables of one of the elements are identical with some input variables of the other element.

A set of recognized elements of a system is usually called the universe of discourse of the system. The name "structure of universe and couplings" is used for the collection of elements with their couplings. I shall use the abbreviation "UC-structure".

Depending on the problem to be solved, the system can be identified by one of the following traits:

1. A chosen set of input variables; a set of output variables; and a set of resolution levels, one defined for each of the variables.
2. A given activity. This includes the variables and the resolution levels specified in (1) and a record of values of the variables within a particular period of time.
3. A given behavior. This includes the variables and the resolution levels specified in (1) and a set of dependencies of the present values of the output variables on the present and/or past values of the input variables and/or the past values of the output variables.
4. An ST-structure, which incorporates (1) and (3), and in addition includes information about internal variables.
5. A UC-structure, which incorporates (1), (3), and (4), and in addition specifies a composition of the behavior and the ST-structure of the system by the behaviors and ST-structures of some elements.

Figure 1 shows a simplified illustration of the relationship between the system concepts introduced above.

In the problem of synthesis, the system is identified by its behavior and we are required to design a UC-structure that (1) produces the given behavior, (2) contains only elements from a specified set of elements, (3) satisfies certain constraints concerning its form, and (4) satisfies best certain specified objective criteria (criteria of "goodness"). The scheme of the problem of synthesis is shown in Figure 2.

The problem of synthesis, which is essentially the problem of an appropriate decomposition of a given complex relation into available simpler relations, does not usually have a unique solution. The stronger the above

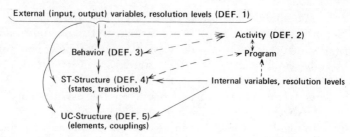

Figure 1.   A simplified illustration of the realtionship between the basic systems concepts.

mentioned requirements (1) through (4) are, the smaller the number of solutions that exist. If the requirements are too strong, there may be no solution of the synthesis at all. This may happen, for instance, if the types of elements that are available (requirement [2]) are not sufficient to compose the behavior or if the objective criteria are too strong or contradictory.

Although the ST-structure is not included in the formulation of the problem of synthesis, its determination usually represents an intermediate step in the procedure from the given behavior to a convenient UC-structure.

The problem of system synthesis has been successfully solved in various branches of engineering. Engineers have developed many procedures for the synthesis of certain types of systems. Some of the most sophisticated procedures of synthesis at the present time are probably those used in computer design. Some of them are strictly algorithmic, and their execution has been fully automated by preparing appropriate computer programs.

Although the problem of system synthesis has not been studied in any significant depth outside of some engineering disciplines, especially electrical or mechanical engineering, it is meaningful in many other disciplines. In

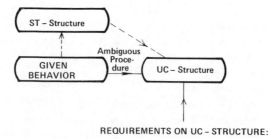

REQUIREMENTS ON UC – STRUCTURE:

1. Types of elements.
2. Constraints concerning couplings.
3. Objective criteria.

Figure 2.   Basic scheme of the problem of synthesis.

economics, for instance, the synthesis of a complex economical unit by some simpler units is certainly meaningful. Similarly, it makes sense to talk about the synthesis of complex health care systems (Smallwood, Murray, Silva, Sondik, and Klainer, 1969), political systems, or social systems. We may even find that the systems synthesis has a meaning in linguistics (the synthesis of artificial languages which find their application, for instance, in computer programming) and in the arts (the synthesis of musical compositions or literary pieces).

In many disciplines we can observe some activities which may correctly be considered as activities whose mission is to solve certain problems of synthesis. However, with the exception of some engineering disciplines, these activities are, as a rule, methodologically at a very low level of sophistication in the following senses:

1. Frequently the problems of synthesis are more or less felt, but not properly formulated.

2. No formal procedures are used to solve these problems, which makes it impossible to use computers for performing them.

3. The principles that are applied are very simple and are based heavily on intuition rather than on a careful study of the problems.

The unsatisfactory situation, as described above, is typical in the so-called soft disciplines such as psychology, the social sciences, or the study of health care systems. They are called soft disciplines because they have not achieved the same maturity as the so-called hard disciplines such as physics or chemistry. It turns out, however, that the soft disciplines deal with problems much more complex and difficult than the hard disciplines. It is an interesting paradox that the best talents have been attracted to the hard disciplines, although the soft disciplines have needed them more. The social scientist encounters much more complex problems than the physicist. One explanation of this unfortunate paradox is that problems in the soft disciplines are so complex that they have not been considered as manageable and, thus, have not promised any challenge to mathematically oriented people.

The situation described above is beginning to change, partly because of the increasing power of our data processing facilities, partly because of social and political pressures, and partly because of changes in education. The general systems methodology plays an important role in this period of change, primarily because it provides the soft disciplines with a basic conceptual framework, basic principles, and methods that can be further elaborated and refined within each individual discipline to serve its needs. A general terminology concerning systems is helpful in improving the communication between disciplines, which is important in solving problems that require interdisciplinary team cooperation.

The general problems of system synthesis are joined by special problems of synthesis of health care systems. The synthesis of a system requires that the following two basic steps be carried out in succession: a well-defined formulation of the specific problem of synthesis should be formed, and the problem should be solved by formal procedure.

The formulation of the problem need not be trivial. It turns out that this very step is frequently the major trouble point, particularly in the soft disciplines. We are usually able to get an intuitive feeling for the problem and have no difficulty discussing it in vague terms. However, when it comes to the precise formulation needed as a solid basis for a formal procedure of synthesis, we often get lost. The precise formulation of the problem requires that the following items be clearly identified: (1) the variables involved and their classification as input, output, and internal variables; (2) the resolution level for each variable under consideration; (3) the behavior that we want the system to perform; (4) the types of elements (subsystems) available to implement the given behavior; (5) various constraints concerning couplings between the elements; and (6) objective criteria.

We may assume, as an illustration, that the object under investigation is a hospital on which we define a simple system with one input variable X and one output variable Y. The input variable X can be defined as arrivals of individual patients, and the output variable Y can be defined as their departures. At the input, each patient is identified by his name and/or other features (age, sex, social security number, type of health, and insurance), which will be denoted by I, and appears in a certain health condition, H_a, which either requires or does not require emergency care. Thus values of the input variable are ordered pairs (I, H_a), when I may contain several items identifying the patient and H_a has two values: emergency care required, e; routine care sufficient, r. The time resolution level of the input variable is based on the maximum expected frequency of arrivals of patients.

Values of the output variable, Y, can be defined by pairs (I, H_d), where I has the same meaning as before (an identification of the patient) and H_d is the health condition of the patient at the time of his departure from the hospital. Clearly, H_d has a different resolution level than H_a. It may contain, for example, four values: no further care is needed, n; home care is sufficient, h; further visits in the hospital are required, v; the patient has died, d.

The behavior of the system is any required dependence of $Y = (I, H_d)$ on $X = (I, H_a)$. Assume that $(I_j H_d)$ is dependent of (I_k, H_a) if $I_j \neq I_k$. If $I_j = I_k = I$, then a set of required conditional probabilities, $P(I, H_d | I, H_a)$, can specify the behavior of the system. An example of such a behavior specification of the system is given in Table 1.

Observe that

$$\sum_d P(I, H_d | I, H_a) = 1$$

must be satisfied for every particular pair (I, H_a). Each of the probabilities $P(I, H_d | I, H_a)$ can be considered as the sum of probabilities $P_t (I, H_d | I, H_a)$ specifying the change that the patient I, who arrived at the hospital with the health condition H_a at time t_1, leaves the hospital in condition H_d at time t_2, where $t = t_2 - t_1$.

Table 1.   An Example of the Behavior of the System Specified in Figure 3.

H_d / I, H_a	$P(I, H_d | I, H_a)$			
	n	h	v	d
I, e	0.4	0.3	0.25	0.05
I, r	0.6	0.2	0.19	0.01

Elements by which the above described behavior can be implemented may be those shown in Figure 3. The behavior of each of these elements must be precisely defined, similar to the way we defined the behavior of the total system. The probabilistic description is again likely to take place in the definition of the behavior of each of these elements. Each of the elements in Figure 3 is itself a complex system that is built up of simpler elements. This illustrates that the synthesis is applicable at various levels of the system hierarchy.

Examples of constraints concerning couplings between the elements can be seen in Figure 3. For instance, the output of the element identified as "arrival classification" must not bypass both the elements identified as "diagnostics" and be coupled directly to the element called "surgery."

The objective criteria for the example discussed may be, for example, the minimum cost, the maximum reliability (the ability to give immediate care to all emergency cases), the minimum average time spent by a patient in the hospital in individual categories, or any of numerous others.

When the problem of the synthesis of a system is precisely formulated, various procedures can be employed to solve it. The procedure which can be used depends significantly on all aspects of the problem. For instance, quite different procedures must be used for continuous as opposed to discrete variables, deterministic versus probabilistic behavior, different objective criteria, strong or loose constraints of couplings, and the like. In the system discussed, the variables are discrete, the behavior is probabilistic, the constraints concerning the couplings are strong, and the objective is, say, to minimize the cost for a given reliability and an average time spent in the hospital for some specified categories of cases. A good procedure under these circumstances may consist of the following steps:

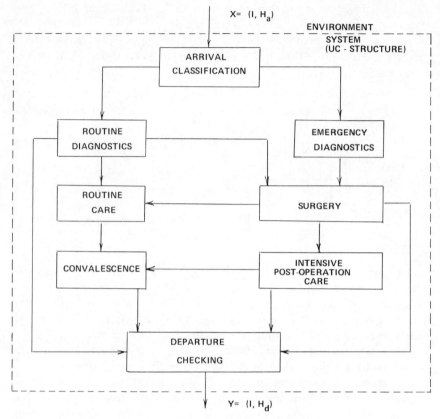

Figure 3.   A simple model of a hospital.

1.   Suggest a basic model of the UC-structure by which the behavior can be implemented. An example is the model shown in Figure 3.
2.   Substitute for the individual elements systems that have been designed before and whose behavior, cost, average time delay, etc., are completely known.
3.   Simulate the UC-structure on a computer for random distributions of values of the input variables and observe the performance of the system.
4.   Change the elements and repeat steps one through three.
5.   Change the model of the UC-structure, if possible, and repeat steps one through four. If it is not possible or not reasonable to modify further the model of the UC-structure, compare the results of the simulation from the standpoint of the prescribed objective criteria and select the best solution.

A significant improvement can be expected in synthesis of health care systems if some health care specialists on various levels become familier with basic systems principles and methods and with some mathematical and computer techniques by which the methods can be implemented.

REFERENCES

Fanshel, S., and Bush, J. W. A health-status index and its application to health-services outcomes. *Operations Research,* 1970, *18,* 1021–1066.

Klir, G. J. *An Approach to General Systems Theory.* New York: Van Nostrand Reinhold, 1969.

Smallwood, R. D., Murray, G. R., Silva, D. D., Sondik, E. J., and Klainer, L. M. A medical service requirements model for health system design. *Proceedings of the Institute of Electrical and Electronics Engineers,* 1969, *52,* 1880–1887.

Concepts and Problems in the Test and Evaluation of Health Systems

Kenyon B. De Greene

Today few, if any, Americans are satisfied with the status of health care, even though the United States spends a greater proportion of its gross national product on health services than does any other country. The United States spends much more on biomedical research than does any other country and has the world's third or fourth highest ratio of professional health personnel to population. Opinions differ widely, however, as to what is wrong and what can be done about it. (Compare, for example, discussions in Galliher, 1969, and Bazell, 1970.) Much of the difficulty derives from the inability to define health itself in a quantitative manner adaptable to different situations; the inability to define the health "system" if, indeed, it is a system at all rather than a congeries of bits and pieces; and, finally, the inability to agree as to the goals of a health system and their quantitative definition. Treatment itself is sometimes an indefinable complex of detection, intuition, diagnosis, and activities.

More specifically, major problems identified in the health field include inadequate and inefficient use of all personnel, facility, and monetary resources; unnecessary hospitalization and surgery; gross variations in insurance coverage; spiraling costs for all aspects of health care, especially hospitalization; duplication of facilities and equipment; poor integration and coordination among physicians, clinics, and hospitals; and generally gross variations in the equality of health care. Solutions involve broadening the personnel base and using new types of specialists, such as paramedical personnel; utilization of additional and new forms of training; introduction of automated equipment and information systems; establishment of integrated programs; and emphasis on prophylaxis over treatment.

This paper will examine some of these problems from the viewpoint of systems evaluation. I shall attempt to establish principles applicable to the entire national health field, the local neighborhood outpatient clinic, or an individual ward within an individual hospital. Important considerations for systems evaluation involve assessment of the effects of one or more of the above or other proposed solutions. Are the effects complementary, indifferent, or even antagonistic? What decisions can be made *before* potentially dangerous policies are implemented?

Implied in the discussion will be such questions as: What can be done to reduce the waste and inefficiency characteristic of much health care? How can we use more effectively those resources we now have, that is, provide more care more efficiently and at less cost? What sorts of new resources are desirable or necessary? How should resources be devoted to medical research for tomorrow as opposed to services for today? What resources should be applied to preventive care and what to curative care? What resources should be applied to service for one age group and what for another? And what sorts of systems concepts and techniques are applicable to the health situation?

COMPLEX SOCIOTECHNICAL AND SOCIAL SYSTEMS

Health systems, which are sociotechnical and social systems, comprise just one member of a complex family involving education, welfare, poverty, insurance, migration, transportation, urban affairs, environment, and scientific and technological advances. Evaluation of these systems individually, but more importantly in an interrelated macrosystems context, is indispensable to the preparation of a social report (U.S. Department of Health, Education, and Welfare, 1969) and to the continued stability, success, and progress of our country. Each of these areas, or subsystems or systems, and others perhaps, interface with the health system and augment it or place stress on it. Because each places demands on the finite monetary, personnel, material, and space resources of the macrosystem we call the nation, health systems experts must develop as meaningful plans for evaluating effectiveness, benefits, costs, trade-offs, and intangibles as are possible. This is true, even if a full-scale test and evaluation program as utilized in military and space systems is impracticable in terms of cost and personnel. This is especially true now that health costs are rising rapidly and are an increasing percentage of income — without, however, demonstrable concomitant improvements in health care.

Sociotechnical and social systems, however, possess attributes that set them apart from the traditional systems with which systems engineering has had to deal. Some of these attributes are that:

1. "Systems" may come into existence with no clear definition of long-term goals.

2. Growth is usually by trial and error.

3. There are usually myriad competitive "subsystems" which typically operate at cross purposes and which may be very difficult to integrate.

4. The systems are nonlinear.

5. The systems are high order (where order refers to the numbers of integrations).

6. The systems are characterized by numerous and interconnected, multiple positive and multiple negative feedback loops operating at different levels; consequently growth, equilibrium, stagnation, and decay may occur contemporaneously.

7. System inputs are usually stochastic.

8. There is usually no clear relationship between "system" and "management."

9. Relationships among designer, purchaser, manager, and user are quite complex; as a result, most analyses and evaluations are biased.

10. System periods may be difficult to decipher and there may be time delays; corrective actions may exacerbate underlying difficulties.

11. Variables may be highly correlated, but true cause-and-effect relationships may be very difficult to decipher.

12. There typically is both a feast of useless data and a famine of necessary data.

13. Long-term and short-term responses to policy or design changes may be diametrically opposite.

14. Solutions may be threatening to the status quo.

Systems science brings to the solution of complex social problems several important concepts and techniques. From systems engineering, Forrester (1964) stresses the concept of a system as a designable entity; the concept of feedback control; the low cost of electronic communication and logic; the substitution of simulation for analytic solutions; and the distinction between policy making and decision making. To these may be added, from the wide variety of presently unreconciled theories and techniques, the theory of open systems; the concepts of sociotechnical systems; the integrated systems engineering-management package associated with the systems development cycle; and the theories and practices of operations research.

SYSTEMS MANAGERIAL CONSIDERATIONS

For purposes of systems evaluation the crucial question is: Does the system do what it is supposed to do? Techniques for the test and evaluation of systems have long been utilized in space programs and by the military (see, for example, Department of the Air Force, 1967; Meister and Rabideau, 1965). Systems evaluation, an integral part of systems development, is just one aspect of a temporal, iterative series of processes involving problem recognition and proposed solution to the identified problems. Test and evaluation of such systems may be nearly as expensive as development. However, the extension of these methods to sociotechnical and social systems presents certain difficulties, especially with relation to the definition, measurement, and control of variables and parameters, to the determination of criteria, and to the availability of data. Nevertheless, many of the large-scale social programs approved by Congress do require an evaluation of the program.

Before proceeding further, I must distinguish between certain terms. "Tests," or "testing," usually refers to the real-time collection of data under a plan and under controlled conditions. "Evaluation" then refers to the a posteriori interpretation of test results. "Systems test and evaluation" can be considered to begin once a given design is initiated and to continue through system operation. In contrast, "systems analysis" can be considered to be antecedent to selection of a given design. Of course, these distinctions are not meant to be restrictive. In most systems, analysis merges into test and evaluation, and comparison and evaluation of alternative design approaches are, by definition, an integral part of systems analysis. Further, "systems" generalizes across the field of systems science; "system" refers to a particular system. (For a further consideration of these definitions and distinctions, see De Greene, 1970a; Department of the Air Force, 1967; Sackman, 1970.) Concepts, problems, and methods considered are generalizable to evaluations whenever in the systems development cycle they are applied.

Effective test and evaluation of systems can be undertaken only within the framework of certain systems concepts. Questions that must be asked and answered as specifically and quantitatively as possible are: What *is* the system under consideration? How can it be defined operationally? What is the system environment and what sort of interactions take place between system and environment? What are the goals and objectives of the system? What constraints restrict the operation of the system? Are the system goals compatible with one another? What hierarchical levels both inside and outside the system must be recognized both in planning for test and evaluation and in interpreting the results? How can test and evaluation results be integrated at successively higher hierarchical levels? What are the main subsystems not only hierarchically but also in the sense of the "social subsystem," the "technological subsystem," the "administrative subsystem," the "user subsystem," and so on? In answering these questions, both system constituents or elements at any one time and their change over time must be considered. As an example, viewing the patient out of the context of his past development and health practices is meaningless.

Answering the above questions in the case of health systems may be quite difficult. The health system is diffuse, and myriad organizations — federal, state, local, and private are founded for administrative, religious, memorial, philanthropic, or purely profit-making motives — are responsible for some aspect of health policy and care. Independent hospitals and health professionals operate on a small scale and discourage coordination and integration of system elements. This leads to both omissions and costly, inefficient duplications of effort, and it confuses many persons in knowing just where and how to enter the health system (U.S. Department of Health, Education, and Welfare, 1969). The evaluation of integrated facilities, equipment, and personnel provides one of the more straightforward applications of systems science to the health field. In other areas a general recognition of health problems may not lead to a practical way to solve them.

Like any serious systems activity, evaluation requires planning. A test and evaluation plan should be prepared indicating the goals of the evaluation; who will do the evaluating; what will be the makeup of the evaluation team; how, when, and where will the evaluation be conducted; what are data requirements and the realistic collection of data; what methodology and subjects are to be utilized; and what are estimated costs and time to be spent? Careful planning will prevent needless duplication of effort.

Test and evaluation should be performed in a realistic operating environment, for example, involving the actual treatment of patients with measurement expressed in terms of performance capabilities. A strictly laboratory approach will produce misleading and perhaps meaningless results. Usually a sequence of evaluations will be necessary, because the system exists in a dynamically changing environment. Typically, it is desirable to perform pilot evaluations as design proceeds so that changes can be made if necessary. Follow-up tests will involve progressively firmer design and successively higher system levels. Also, the results of a new design may not be available for many years.

Ideally, evaluation should involve all the system constituents: equipment, facilities, communications, computer programs, handbooks and manuals, and personnel, including personnel skill and manning levels and training. Thus, there is considerable evidence that technology (Lehan, 1969) and manpower (Wolfe, Iskander, and Raffin, 1969) are not being utilized effectively. Some of the system constituents can be evaluated in isolation initially, but eventually total systems testing and evaluation should involve the realistic situation of each interacting with the others. Traditional considerations of hardware and software reliability, safety, maintainability, logistics, and human engineering may find application here. An example of this kind of study is provided by Flagle (1970), who looks at criteria for evaluation of a medical information system. These criteria might be such things as completeness, timeliness, reliability, operability, and cost, each expressed as measures of effectiveness relative to various medical system functions and operations like diagnosis, patient care, research, and administration. Such measures of effectiveness must be subject to careful trade-off study; for example, reliability can be increased by added redundancy, but this also increases costs. This kind of evaluation is typical in the health field. In rare cases the system will be totally new; much more likely is the situation wherein, say, a new computer is added to an already existing facility. Such incremental growth is typical of most health systems.

Another type of evaluation, particularly important at higher systems hierarchical levels, involves determination of the effects of different policies and programs. Forrester (1969), using a complex urban model and computer simulation interrelating housing, industry, and people, has studied growth and decay patterns in a city over a 250-year period. Normal aging processes resulted in economic and otherwise undesirable combinations. Various urban management programs to ameliorate these problems, such as job aid, additional training, low-cost housing, and tax expenditures for welfare and education, could be simulated and the effects studied. The results of these programs ranged from no effect to detrimental. This can be explained in terms of counterintuitive behavior, that is, intuitively sound corrective decisions and actions may actually aggrevate a condition, which calls for more of the presumed corrective action which worsens the situation even further. An example of such a study applied to the health field might entail the effects of adding substantial manpower and dollars to the present "system" of health care. It is of paramount importance to evaluate any such proposed solutions or designs before they are implemented.

Evaluation should involve not only the usual operating situation, but also the effects of contingencies and critical stresses upon the system. How do these affect its capabilities and limitations? What are the effects of overloads? Of disasters? Of severe environments? Further, evaluation should be in terms of the ongoing growth of the system, realizing that the system is not static.

Evaluation must be viewed within both longitudinal and cross-sectional contexts. Failure to realize this may lead to an optimum solution to the wrong problems. For example, in the Netherlands, a country having one of the lowest infant mortality rates in the world, nearly all expectant mothers receive prenatal care but a sizable proportion of babies are delivered at home rather than in a hospital. In the United States, on the other hand, one-third to one-half of the urban mothers who deliver at public hospitals have received no prenatal care (U.S. Department of Health, Education, and Welfare, 1969). Almost all American babies are now delivered at hospitals, even though "the running of the obstetrical facilities has become a losing proposition in hospitals in many parts of the country" (Wolfe et al., 1969, p. 369). This last situation leaves the health systems professional with the problem of doing the best he can under quite constrained conditions to optimize use of personnel and other resources.

Test and evaluation results may be meaningless, and the system itself a failure, if the users are not brought in early. In this context, users include both the operators (such as physicians, nurses, and laboratory technicians) and administrators, and the public which is served. The best of intentions and new technological developments have gone awry because of failure to recognize the importance of operator training, of individual perceptions as to job threat, of resistance to things imposed from outside, and so on. Thus, technological advances, supposedly good, stress organizations and systems. How does the system keep up with such changes? What interrelationships exist? What are the effects on job role, self-concept, training, and education? These are some of the questions that must be posed and answered.

If the effects of the proposed changes to the system are not determined before they are actually made, the new technological and operational improvement may actually lead to performance deterioration. This feature of the evaluation of systems involves the prediction of the effect on the system of the change made to any particular system element, or, contrariwise, the prediction of the effect on an element of changes made elsewhere in the system. Sensitivity analysis and sensitivity testing are terms used to describe how assumptions, based on uncertainty, influence an analysis or evaluation; for example, how a criterion of systems effectiveness is affected by changes in a variable(s) that determine the value of that criterion.

Finally, a concept important to systems test and evaluation and one long practiced relative to military, space, and some commercial systems is that of an experimental system, that is, a hospital (National Advisory Commission on Health Manpower, 1967) or an "operational testing laboratory" (Horvath, 1966), which might be, for example, a small community. Here it would be possible to evaluate new procedures, new regulations, new methods of treatment, new patient-physician-nurse relationships, and new insurance methods and modify them as necessary before applying them to the city, state, or nation. Results that are dynamically fed back and compared with prognosis provide the basis of a cybernetic control system. For example, previous practices of prevention or treatment, and of costs and benefits or effectiveness, can be reevaluated and recalculated. A particularly good example of the application of systems techniques is the USPHS Health Evaluation Center (automated multiphasic health testing system: Hsieh, 1969; U.S. Department of Health, Education, and Welfare, 1970), which provides the capability to evaluate the usefulness of systems engineering methods to the solution of health problems.

The future undoubtedly will witness the emergence of health systems teams comprised of experts in systems science (general), behavioral and social science, biological science, economics, medicine including epidemiology and nursing, operations research and statistics, engineering, computer science, management, and also the clients or patients. One of the main roles of such teams will involve defining exactly the operational configuration of the health system and determining the extent that it does what it is supposed to do.

METHODOLOGY

The simplest kinds of evaluations are those which resemble traditional scientific laboratory experiments. The usual statistical methods are applicable in such cases. At another hierarchical level, algorithmic operations research techniques (analytic models) may be applicable. The most involved, system-oriented evaluations contain so many parameters as to make mathematical modeling impossible; heuristic models, computer simulations, and games may be appropriate at this level. In addition, results must often be interpreted in terms of the consensus of a group of professional health experts. I shall consider these methods later. What is considered as a system level is relative. However, for our purposes here, systems hierarchy can be represented as follows:

1. Macrosystem—a national or regional composite of several community health systems together with the necessary coordination and integrative organization and management structure.
2. System—a composite of hospitals, clinics, laboratories, drug and equipment supplies, and physicians' offices, and their accompanying organization and management structure, dedicated to the health services of a community (usually a town or city).
3. Subsystem—an individual hospital, clinic or laboratory, or group medical practice.
4. Sub-subsystem or module—a physical plant, a computer assemblage, a software package, the equipment in an automated laboratory, etc.

The lowest systems' hierarchical constituents—components, units, and parts—will not be considered here (see De Greene, 1970b).

Methodology should be of a general nature and should be applicable to different populations, health resources, political organizations, economic conditions, climates, and so forth, yet applicable within specific organizations as needed. Allocation decisions typically must be made both within specific programs and among different programs. The general-purpose approach should be capable of being tailored to any systems' hierarchical level and to such broad areas as environmental control, accident reduction, mental and nonmental illness, institutions, treatments, insurance, education, and vector control. Hopefully, the decision maker would then be able to compare, say, the results of an accident-reduction program with that of an emergency-treatment program.

The heart of systems test and evaluation lies in the criterion problem, whether we refer to performance effectiveness, benefits, or costs. Precise measures of benefit, effectiveness, and improvement may be elusive. Even cost estimation tends to be complex; and it is necessary to differentiate among fixed, variable, past, future, direct, and indirect costs and, in addition, to specify who bears the costs — the government, the community, or the individual. Further, what one considers a criterion or a constraint, a measure of cost or a measure of benefit, may be a relative thing. Some representative cost, effectiveness, and benefit criteria are shown in Table 1.

A LOOK AT EFFECTIVENESS

The criteria for effective performance of a health system may be diffuse, involving intangible and qualitative considerations. Indeed, what *is* health? A multidimensional scale is almost always necessary when measuring health system performance. This problem is particularly true with regard to mental health, where incidences are unknown, psychiatric diagnoses imprecise, and behavioral manifestations constantly changing with culture. These factors may make comparative evaluations very difficult. Also, effectiveness of mental health programs may be as much a function of cultural and social factors as the specific mode of treatment.

Measures of effectiveness of a mental hospital might be the smallest ratio of time a patient is in the hospital to the time he is in the community, or the smallest number of beds necessary to serve a standard number of the population (Ullmann, 1967). Likewise, size of a mental hospital appears to be negatively associated with effectiveness and size of staff positively associated with effectiveness.

Packer (1968) defines at least three ways to measure effectiveness; this, in turn, affects how decisions are made and who makes the decisions. These involve, first, assignment of a definite numerical value to each effectiveness result; second, the simple ordering of each effectiveness result so as to rank each as better, equal, or worse than the others; and, third, the presentation to the policy maker of the complete effectiveness results of alternate system con--figurations and letting him make his own choice.

Table 1. Representative Criteria* in the Evaluation of Health Systems

Cost Criteria	Effectiveness Criteria	Benefit Criteria
Direct fixed costs, e.g., interest charges associated with past construction expenditures	Increase in national life expectancy at birth and at given subsequent ages	Individual salary increase resulting from cure of a certain disease
Direct variable costs, e.g., costs of materials, supplies, personnel, and equipment required by a growing population	Number of patients treated per "facility" per time period	Tax returns to the community resulting from elimination of a certain disease
	Number of persons practicing prophylactic care within a given geographic area	Increased productivity in man-years and equivalent salary resulting from decrease in accidents
Direct past costs, e.g., costs of a facility the bonds of which have been paid off	Increase in national expectancy of healthy life (free from bed disability or institutionalization)	
Direct future costs, e.g., personnel salaries and supply increases associated with inflation	Percentage of discharged patients who do not require subsequent treatment	
Implied or indirect costs, e.g., potential earnings lost because of mental illness	Reduction in incidence of a given disease within a given geographic area	
	Reduction in infant mortality within a given geographic area	
	Accessibility in miles or minutes to a given portion of a population of a given type of health care	
	Availability in terms of no-cost restrictions of health care to an entire population	
	Subjective and personal satisfaction of a patient with health care	
	Quality of care as determined by a panel of experts	
	Capacity in terms of beds of a given facility	
	Degree of utilization of a given facility	
	Variety of health services offered by a given facility	
	Patients not turned away by a given facility	
	Waiting time for a given service in a given facility	

* Associated constraints are implied in several examples.

Some workers feel it may be easier to measure system ineffectiveness, the complement of system effectiveness. This is sometimes defined as the absence of ill health. The ability to define effectiveness or ineffectiveness of course differs among disease entities. Sometimes the difficulty in defining health is approached by utilizing a "health index" based on mortality or morbidity rates, but an operationally useful index has not yet been developed. One approach is that of Packer (1968), which assigns the value 0 to the state of death, 1 to bed confinement, 2 to a disability resulting in a major restriction of activity, 3 to a disability resulting in a minor restriction of activity, and 4 to good health. The health index is the sum of the products of one of these weights and the respective probability of a person's being in that state of health. Such an index seems to be an improvement over purely monetary approaches based on real and implied earnings.

A LOOK AT COSTS

When we attempt to assay the costs of illness, we must consider all sorts of indirect costs to relatives and to the community. These may be very difficult to quantify. As far as the ill individual himself is concerned, costs can be expressed in terms of loss of earnings, loss of taxes, cost of sustenance and care inside and outside an institution, workless days, reduction in productivity, and the like. However, *exact* estimates of costs remain unlikely because of the criterion problem; this is especially true when we try to define mental illness and the scope of the mentally ill. Cost estimates are invariably both over-simplified and conservative. How do we include such intangible, subjective, and qualitative considerations as loss of self-esteem and degradation of the very quality of life?

A more precise definition of the costs of mental illness with actual dollar examples follows. Conley, Conwell, and Arrill (1967), using 1966 data, estimated the total cost of mental illness in the United States at about $20 billion annually, almost 3 percent of the gross national product. Cost was defined as the loss of well-being suffered by society as a result of disease.

An especially thorough analysis of the economic cost of disease has been made by Hallen, Harris, and Alhadeff (1968). They determined the direct costs of personal services (hospital care, nursing home care, professional help, and drugs) and nonpersonal services (research, training, other health services, construction, and insurance), and the indirect costs of morbidity and mortality associated with kidney disease. Total economic costs for 2964 were estimated at abot $3.6 billion, of which $1.2 billion was attributed to direct costs and $2.4 billion to indirect costs.

COST EFFECTIVENESS AND BENEFIT/COST

In modern systems management practice, costs are usually related to "effectiveness" or to "benefits" as cost/effectiveness or benefit/cost (ratios). These concepts differ somewhat, and not all users agree as to definitions. Usually, cost-benefit relates the dollar costs and dollar value of ostensible benefits of some effort. Cost-effectiveness expresses costs in dollars, but effectiveness is usually expressed in other terms, for example, number of patients cured by a given treatment. Classical applications of cost-effectiveness analysis to health are those of Papanicolaou smear tests for cervical cancer and X-ray screening for tuberculosis (see Horvath, 1967).

Benefit-cost analysis assumes that both benefits and costs will be expressed in the same units. But how does one value in dollars lives saved and illness avoided? Monetary losses associated with morbidity and mortality include both indirect costs (decrease or loss in economic productivity) and direct costs (of capital investment in facilities and of detection, treatment, and rehabilitation). Assumptions and methodology are imperfect because of difficulties in assessing productivity, the value of women, the value of the aged, and intangibles. Further, future costs and benefits are usually discounted to present values, and economists do not agree on which values to use and how to use them.

Benefit-cost analysis involves calculating the ratio of benefits to costs, which provides a deceivingly simple means of comparing and evaluating alternative programs. A benefit/cost ratio, desirably equal to more than one, can be viewed as an important criterion in policy making.

A particularly thorough benefit-cost analysis has been made by LeSourd, Fogel, and Johnston (1968). This study illustrates not only methodological difficulties associated with estimating and measuring benefits and costs, but also that assumptions regarding infections and epidemiological factors in disease may be deficient. Medical research may not yet have provided the answers.

LeSourd and associates measured benefits as the discounted value to lost income avoided due to prevention and treatment. Costs were measured as federal expenditure per person screened in preventive programs, and as annual federal cost of treatment for patients requiring it. Benefits and costs were obtained by cohort analysis, that is, the study of a group of individuals with one or more characteristics in common. Table 2 illustrates some of their results from a streptococcus screening program for children at a cohort size of 100,000.

Obviously, such measures may be hard to make. Often all one can say is that benefits appear to exceed costs. Benefit-cost analysis typically is specific

Table 2.   Example of Benefit-cost Analysis

Program Alternative	Benefits	Costs	Benefit /Cost
Sponsor central bacteriology laboratory and perform all 100,000 cultures	$2,929,000	$200,000	14.6/1
Use existing facility already screening for Group A Streptococci	$2,929,000	$75,000	39.2/1

Reprinted from D. A. LeSourd, M. E. Fogel, and D. R. Johnston, *Benefit-cost Analysis of Kidney Diseases Programs* (USPHS Pub. No. 1941, U.S. Government Printing Office, 1968).

to given patients and methods of treatment. It may be of limited value in determining optimum mix of health programs; for one reason, because marginal benefits cannot be measured well. As mentioned above, it is usually necessary to rely on the judgments of health experts in such cases.

OPERATIONS RESEARCH

Questions typically asked by the health system practitioner could be: How much time or money should be spent looking for a cure as opposed to extending present treatment concepts? What is the optimum number of patients per physician under given conditions of time and budget? What is the best mixture of treatments, given restrictions on staff? Is a given group of patients able to benefit from a given type of treatment administered for a given amount of time and, if so, at what cost? Resolution of such questions is all-important in times of conflicting demands on limited resources. For example, if a group of patients is not benefiting from available treatment, it might be decided to expend resources otherwise in an environment of limited funds and a growing number of patients. Important alternatives thus are both prevention versus nonprevention, prevention versus treatment, and one form of treatment versus other(s). Another complication is that the accessibility of the patient differs in each case, and thus prevention may be a lifetime process.

Answers to the above and similar questions have been provided by the body of theory, approach, and technique called operations research. A summary of problems and applications of operations research is given in Table 3, and a few interpretations follow.

Operations research has been applied to health problems in general for about two decades (Flagle, 1962), although some areas — for instance, mental

Table 3. Summary of Problems and Applications of Operations Research for Use in the Health Field

Representative problem area(s)	O.R. technique	Typical measurement(s) and parameters	Frequently used
Allocation of (mixes of) costly or limited resources, e.g., beds, physicians, treatments, diets, or drugs, to different kinds of patients	Linear programming	Numbers (of beds, patients, physicians), types of and probabilities of needs for treatments, concentrations of drugs or dietary constituents, dollars, times	Yes
Conflict situations viewing "nature" (of a disease) as the opponent and decision to treat or not to treat the disease	Game theory	Numbers of patients diagnosed, cured, died, etc., probability of having a disease, dollars	No
Stocking of usable, replenishable, spoilable, or seldom used items, e.g., drugs and vaccines	Inventory theory	Amounts of item, times, dollars	Yes
Random arrival and flow of patients through a health "facility"; waiting of patients for treatment or facilities for patients; scheduling of resources	Queuing theory	Numbers of patients, times, number of treatment "facilities," dollars	Yes
Costs of preventing an equipment failure or personnel loss versus cost experienced through such failure or loss	Replacement	Dollars, times, numbers of equipment or personnel	No

Limited resources to detect a disease, type of blood, donor, or need for a satellite location	Search	Numbers of patients, groups of blood, etc., probability of having a disease, times, dollars	No
Prediction of staffing needs, service, health, or disease from observed variables	Multiple regression analysis	Time spent on various aspects of treatment	Yes
States of occupancy for hospital beds over time; states of health, transition from one state to another, and demands for treatment in each state	Markov processes	Times, rates, probabilities of future bed occupancy; probabilities of being in a certain state of health; dollars	No
Transition from one health state to another; computer-based medical diagnosis	Behavioral decision theory	Probability of transitioning from one health state to another; physician subjective probability estimates about symptom-disease relationships	No

This table is partially based on texts in Gustafson, Edwards, Phillips, and Slack, 1969; Halpert, Horvath, and Young, 1970; and Milly and Pocinki, 1970.

health (Halpert, Horvath, and Young, 1970) — have seen few applications. Three observations can be made about the results. First, findings, essentially all at the sub-subsystem, subsystem, or greatly oversimplified system levels, may not integrate into a meaningful description of the real-world system. Second, most of the approaches tend to treat the individual in the aggregate and, hence, depersonalize him. Third, these studies typically deal with system parameters that most resemble those in classical inventory, allocation, and queuing problems. Parameters such as arrival times, time spent with a physician, or bed occupancy may or may not bear a close or direct relation to health. Perhaps these features, the most immediately recognizable and quantifiable, are incidental to health.

It appears that operations research techniques will differ in their usefulness. This might depend, for one thing, on the disparity between the assumptions underlying an operations research application and other systems theory. Systems theory — for example, open systems theory — might predict that health, disease, and recovery are functions of environmental factors such as family and peer group relations, fear and hope, and self-image as much as of specific trauma or infections. Assuming or simplifying these things out of existence, and concentrating on times spent with the physician, time spent in bed, and so forth may lead to attractive but useless conclusions. Further, an understanding and modeling of present user perceptions of need and medical care utilization patterns may distort understanding of the fundamental nature of the problem. Operations research techniques should see their greatest usefulness in such applications as the reduction of redundant facilities and equipment, the identification of lack of standardization of procedures and lack of consistency in standards among hospitals and clinics, and the development of optimum procedures for stocking of supplies and equipment spare parts.

HEURISTIC OR SIMULATION MODELING

As system complexity increases, it becomes more necessary to simplify models. Eventually the model may bear little or no relationship to the real world, and both its descriptive and predicitive capabilities may approach zero if one has not chosen the key variables well. Key variables are those which, if changed, would exert the greatest effect on the system.

In the ideal sense, manipulating the model would involve changing the inputs, for instance, personnel numbers or budget, and calculating the outputs, for example, reduction in specific morbidity or in infant mortality. Things are seldom this simple.

In almost all cases dealing with total health systems, it will be impossible to find optimum solutions, and the search must be directed toward finding a good solution. Another way of saying this is that the algorithmic, analytical, numerical, or symbolic models of classical operations research do not find meaningful application to the solution of problems of complex systems. Rather, here the researcher must employ heuristic or simulation models. There is, however, no absolute distinction between these two classes of models; they grade into one another.

CASE EXAMPLES

Several research examples will illustrate different approaches to the evaluation of health systems. The first is classical experimentation at a lower hierarchic level. This approach involves the usual problems of selection of subjects, sampling, control, and statistical analysis. Watson and Fulton (1967) provide a useful study, an evaluation, in terms of costs and time, and the effects of adding extra staff to a program for treating the psychiatric-medically infirm, largely an elderly group. Two 36-bed wards were compared. The experimental-ward patients were assigned extra staff over an 18-month period; this amounted to an equivalent of $1.89 each day extra per patient. Criteria of improvement were rating-scale values for eating, dressing, bathing, elimination, and ambulation. Results indicated a decrease in dependence on the staff for care, suggesting that expenditure for this elderly group is justified.

A second type of analysis is a system comparison with analysis. Ullmann (1967), in studying individual institutional differences among psychiatric hospitals, evaluated the effectiveness of mental hospitals in terms of organizational and bureaucratic structure and function. He found that both small size and high staff/patient ratio were significantly associated with criteria of effectiveness, but each affected different measures of hospital effectiveness.

Evaluative criteria stressed mainly the return of patients to the community. The measures of effectiveness were operationally defined as: the percentage of newly admitted schizophrenics returned to the community for a period of at least 90 days within 274 calendar days after admission (a measure of early release); and the percentage of patients remaining in the hospital for two years or more (a measure of turnover).

Ullmann performed a functional analysis of behavior involving both patients and employees. He viewed the behaviors of both as a function of actions reinforced by the social environment as each perceived it. He found that reinforcement practices in hospitals fostered custodial care and the displacement of such goals as return of patients to the community. Improvements in effectiveness apparently can be made by administrative and organizational changes which, with the exception of increasing the number of treatment personnel per patient, would not drastically increase expenditures. For example, the percentage of daily expenditures associated with treatment, as opposed to food and shelter, was positively correlated with the measures of effectiveness. A family-care program was also highly correlated in a positive direction with both measures of effectiveness.

Another form of analysis is cybernetic sampling and modeling of operational effectiveness. A basic criterion of the effectiveness of any health system ultimately is user satisfaction. The system will still be a failure, in spite of the newest facilities, the maximum ratio of highly qualified staff to patients, and the most sophisticated computer capabilities, if there is a large gap between the actual system output and what clients and patients expect and perceive that output to be. User feedback is thus an essential feature of total systems evaluation. The requirement for dynamic information as to the disparity between actual and desired performance thus defines a cybernetic system. Dynamic sampling and evaluation are also essential if the health system is to predict incipient trends and modify its goals with changes over time with, for example, the changing age or educational level of the catchment area. An organizational framework would have to be devised to obtain, process, and evaluate relevant spontaneous and requested oral, written, and behavioral information. Krendal's (1970) study of spontaneous citizen complaints to the Philadelphia Mayor's Office of Information and Complaints and cybernetic modeling of response times is a step in this direction.

Modeling and simulation in the development of a health care delivery system is a fourth possible mode of analysis. This deals with modeling and simulation of a community health care delivery system and includes the study of alternative designs (Milly and Pocinki, 1970). This effort is still under development. The alternative designs may range from minor variations within the existing system, such as personnel or training changes, to a completely new system. The alternatives, expressed quantitatively, are to be evaluated against a comparative baseline, for example, a present health care delivery system. The health care delivery system and its associated operations are viewed in terms of supply and demand for services, where supply can be expressed in terms of personnel, facilities, equipment, and costs.

Inasmuch as the real world is characterized by numerous interactions of variables that can only be described in probabilistic terms, these authors conclude that the best evaluation tool is a set of general models that both simulate the details of the real world (community and its health care delivery system) and retain its random nature. The total system model includes a community population model, a health care delivery system operational performance model, a cost model, and a financial model. Given inputs and outputs characterize each model.

Basic to the simulation of health care systems, according to this approach, are the concepts of the "case-type" and "service-treatment station." The case-type (a group of like individuals, for instance, white males, aged 35–44 with diagnoses of coronary heart disease) is the basic element of the population served by the health care delivery system and is defined parametrically in terms of individual needs for health services, demand for and utilization of these services, and related economic considerations. The service-treatment station is the basic element of the health care delivery service and consists of personnel, equipment, and supplies. A facility consists of several service-treatment stations, and a system may comprise several facilities. Statistical, mathematical, and logical relationships portray the interactions of these basic elements functionally and the utilization of resources either in real-world or hypothetical systems. Model building involves determining the numbers of each basic element, the number of parameters (age, sex, family income, and diagnosis) necessary to define each element and also the development of functional relationships. The delivery system has a number of service-treatment stations where given quantitative functions or services are performed. A statistical distribution of many features of treatment is made for each case-type arriving from the population model. This permits the specification of medical management in advance and still allows for random variations. Operational performance evaluation and an overall system performance evaluation, including financial model outputs, can be undertaken. Representative evaluation criteria include patient satisfaction, expressable quantitatively as a measure of performance by, for example, waiting time; provider satisfaction, expressable, for example, by unscheduled duty hours; and accessibility, expressable, for example, by time spent waiting to enter the system. The overall model can be used to test the impact of a new technological development or a new class of health personnel. Use, of course, involves changing given inputs and looking for changes in associated outputs.

Systems theories and techniques have been applied to the health field for about two decades. Today, because of the great interest in "systems" and because of the tremendous social and economic difficulties confronting the nation, the rate of application is much greater than in years past. Applications of systems "know-how" have had and will continue to have an impact on the health field, because their potential is only beginning to be felt.

It is well to bear in mind the words of a pioneer in this field, Young (1969).

Obviously, there is a pressing need for solutions; unfortunately, I see no easy solutions. And I say this after having been involved in the application of operations research to health problems for over ten years. Unlike other speakers who have attempted to compare the health services with other kinds of organizations in seeking solutions, I find that such solutions are largely irrelevant — that they ignore the profound nature of the problems facing us . . . .

I have often been asked the question: show me one instance where the application of O.R. has initiated or led to a *major* decision in the health services — and I am hard pressed to do so (pp. 395-396).

REFERENCES

Bazell, R. J. Health care: AMA white paper offers traditional solutions. *Science,* 1970, *170* (3964), 1287-1288.

Conley, R. W., Conwell, M., and Arrill, M. B. An approach to measuring the cost of mental illness. *American Journal of Psychiatry,* 1967, *124,* 755-762.

De Greene, K. B. Systems analysis techniques. In K. B. De Greene (ed.), *Systems Psychology.* New York: McGraw-Hill, 1970.(a)

De Greene, K. B. Systems and psychology. In K. B. De Greene (ed.), *Systems Psychology.* New York: McGraw-Hill, 1970.(b)

Department of the Air Force. *Test and Evaluation of Systems, Subsystems, and Equipment.* (AFR 80-14) Washington, D.C., 1967.

Flagle, C. D. Evaluation techniques for medical information systems. *Computers and Biochemical Research,* 1970, *3,* 407-414.

Flagle, C. D. Operations research in the health services. *Operations Research,* 1962, *10,* 591-603.

Forrester, J. W. Common foundations underlying engineering and management. *IEEE Spectrum,* 1964, *1* (9), 66-77.

Forrester, J. W. *Urban Dynamics.* Cambridge, Mass.: M.I.T. Press, 1969.

Galliher, H. P. Quantitative methodology for medical management. In G. K. Chacko (ed.), *The Recognition of Systems in Health Services.* Arlington, Va.: Operations Research Society of America, Health Applications Section, 1969.

Gustafson, D. H., Edwards, W., Phillips, L. D., and Slack, W. V. Subjective probabilities in medical diagnosis. *IEEE Transactions on Man-Machine Systems,* 1969, *MMS-10* (3), 61-65.

Hallan, J. B., Harris, S. H. B., III, and Alhadeff, A. V. The economic cost of kidney disease and related diseases of the urinary system. (USPHS Pub. No. 1940) Washington, D.C.: U.S. Government Printing Office, 1968.

Halpert, H. P., Horvath, W. J., and Young, J. P. *An Administrator's Handbook on the Application of Operations Research to the Management of Mental Health Systems.* (National Clearinghouse for Mental Health Information Pub. No. 1003) Washington, D.C.: U.S. Government Printing Office, 1970.

Horvath, W. J. Operations research in medical and hospital practice. In P. M. Morse and L. W. Bacon (eds.), *Operations Research for. Public Systems.* Cambridge, Mass.: M.I.T. Press, 1967.

Horvath, W. J. The systems approach to the national health problem. *Management Science,* 1966, *12,* B-391 to B-395.

Hsieh, R. K. C. System design of health service system in the public health service hospitals. Paper presented at the System 70 Conference, Long Beach, California, November, 1969.

Krendel, E. S. Systems engineering and the quality of urban life: A case study of citizen complaints as social indicators. Unpublished paper, Department of Operations Research, University of Pennsylvania, 1970.

Lehan, F. W. Impact of new technologies on health services. In G. K. Chacko (ed.), *The Recognition of Systems in Health Services*. Arlington, Va.: Operations Research Society of America, Health Applications Section, 1969.

LeSourd, D. A., Fogel, M. E., and Johnston, D. R. *Benefit-cost Analysis of Kidney Disease Programs*. (USPHS Pub. No. 1941) Washington, D.C.: U.S. Government Printing Office, 1968.

Meister, D., and Rabideau, G. F. *Human Factors Evaluation in System Development*. New York: Wiley, 1965.

Milly, G. H., and Pocinki, L. S. *A Computer Simulation Model for Evaluation of the Health Care Delivery System*. (Health Service and Mental Health Administration Report No. HE 70-15) Washington, D.C.: U.S. Government Printing Office, 1970.

National Advisory Commission on Health Manpower. *Report of National Advisory Commission on Health Manpower*. Vol. 2. Washington, D.C.: U.S. Government Printing Office, 1967.

Packer, A. H. Applying cost-effectiveness concepts to the community health system. *Operations Research*, 1968, *16* (2), 227-253.

Sackman, H. Systems test and evaluation. In K. B. De Greene (ed.), *Systems Psychology*. New York: McGraw-Hill, 1970.

Ullmann, L. P. *Institution and Outcome: A Comparative Study of Psychiatric Hospitals*. Oxford: Pergamon Press, 1967.

U.S. Department of Health, Education, and Welfare. *The Health Evaluation Center at the USPHS Hospital*. Washington, D.C.: U.S. Government Printing Office, 1970.

U.S. Department of Health, Education, and Welfare. *Toward a Social Report*. Washington, D.C.: U.S. Government Printing Office, 1969.

Watson, C. G., and Fulton, J. R. Treatment potential of the psychiatric-medically infirm. Part I. Self-care Independence. *Journal of Gerontology*, 1967, *22*, 449-455.

Wolfe, H. B., Iskander, M., and Raffin, T. A study of obstetrical facilities. In G. K. Chacko (ed.), *The Recognition of Systems in Health Services*. Arlington, Va.: Operations Research Society of America, Health Applications Section, 1969.

Young, J. P. No easy solutions. In G. K. Chacko (ed.), *The Recognition of Systems in Health Services*. Arlington, Va.: Operations Research Society of America, Health Applications Section, 1969.

Research Directions

The presentations in this volume suggest a broad spectrum of health care problems to which systems theory and methodology may be applied. An overview of the implications for research provides interesting and intriguing ideas.

One major area for study is the decision-making subsystem. Questions were raised by the contributors to this volume about the kinds of decisions health care practitioners are required to make and the nature of alternate courses of action that are available. The critical nature of many health care decisions gives rise to questions about the identification of information needed to make a choice from several alternate courses of action. Technological advances and tremendous resource allocation permit dramatic impact on patient outcomes in selected instances. Criteria for the selection of the individuals to be the recipients of this kind of health care have not been established. The preparation of the health care practitioners to make decisions about patient selection has not been defined, nor have any criteria been identified to assist them. That a great range of difference in decision-making ability as it relates to health care exists is indisputable. There is a need to identify the variables contributing to the decision-making ability. Among those presenting intriguing areas of study are psychological variables, organizational setting variables, and informational variables. The degree to which the recipient of care participates in the decision-making process needs to be determined much more clearly. Whether such participation should be expanded and encouraged demands extended study. The effect of expanded participation of the recipient of care in the decision-making process can have great impact on the health care system; the direction and degree of that impact need to be studied.

The nature of the information that is essential in the performance of each health profession is another area for study. The effect of the quality and quantity of the information on the components of the system ought to be measured. Studies are needed to develop information as a major resource in medical intervention. Methods need to be pursued to make information easily accessible and useful in the health care system. Facilitation of information flow needs to be studied, using whatever technological innovations that are possible. Intervening information links need further study. Methods of measuring the effects of communication on behavior is an area of study that has hardly been touched. Criteria for effective communication in the hospital setting need to be identified. Varying numbers of health care workers are involved in information gathering and processing; the effects of these activities on patient care and patient satisfaction with care ought to be measurable.

Implications for research in the area of education of health care workers were suggested by a number of contributors. Tenney says that justification for many teaching methodologies is nothing more than tradition; research is needed to identify the impact of present methods and to suggest innovative approaches. Studies need to be done on cost-benefit ratios related to continuing education, especially for use when comparing the systems approach in continuing education to traditional methods. The use of systems research methods provides tremendous opportunity to study needs of continuing education in terms of having practitioners perform in accordance with agreed-upon criteria of patient outcomes. Zagornik suggested that the systems approach should be applied in the building of coordinated curricula for health professional schools. Study is needed to determine the type of education required to prepare people in the health professions for various roles (e.g., administrator, supervisor, and practitioner). Identification of the determinants of the length of educational preparation time for various health professionals is necessary. And within the health professions, educational requirements for different levels of practice need to be delineated.

While research efforts are being directed toward analysis of clinical tasks to determine how they can best be allocated to the multidisciplinary team, there is a great need for further study in this area. Clearly, there is a need to determine the appropriate allocation of health care resources to emergency and routine situations. It is entirely possible that the health care recipient and those in his community of acquaintances might take more responsibility for some care functions. Thus, it becomes necessary to establish criteria for such allocations. Tenney discussed the need to specify objectives for the physician in his professional function. The boundaries of professional function must be defined, and, if overlapping boundaries are found, potential conflicts must be identified and alleviated. Resource allocation and utilization have an impact on quality of care, a nebulous concept whose quantification most certainly could provide research problems for a lifetime.

Research on professional performance including work load and staffing patterns is also indicated. The organizational and managerial factors that affect performance, work load, and staffing patterns need to be identified.

Different research methodologies have been used in studying various aspects of the health care delivery system. It would seem, however, that an important distinction needs to be made with respect to methodology. In applying systems theory, the researcher must carefully consider the advantages and disadvantages of utilizing either a descriptive or a normative approach. These are not necessarily antithetical, but they certainly emphasize different purposes for conducting research.

Cost-benefit research in health care is another approach that provides a fertile area of study. The problem of criteria to be established for cost-benefit effectiveness has not been conclusively studied.

The effects of patient payment and physician remuneration on medical practice are interesting and controversial variables that warrant examination. Research in health system accessibility and degree of utilization also suggests examination of payment practices. Tenney's paper suggests great need in the medical intervention subsystem to study procedural rules and constraints relative to practice and education; these same areas should be studied in other health professions. Furthermore, we must not lose sight of the interactions of the subsystems in health care delivery in terms of facilitating and inhibiting effects.

The systems approach as a heuristic device for the development of theory is used by investigators from different disciplinary backgrounds. King suggests the feasibility of the use of systems research in studying nursing activity with all of its complexities. There is the possibility of determining the essence of the nursing process through studies of nursing function which consider all the variables involved in providing nursing care. There is need for testing and identifying fundamental concepts to provide the basis for sound theories concerning nursing and health care.

Hearn believes that system studies of social work could lead to fundamental theories applicable at any level of intervention. Vigorous application of systems theory should determine the possibility of developing a social work theory that has meaning for intervention at the level of the individual, the family, the peer group, and the community. The activities of social workers in helping the client regulate his own boundaries as to their degree of openness or closedness need to be identified. Conceptualizations of social work function must be tested empirically.

The boundaries of the subsystems and variables of transaction between them and between the total system of health care and its environment suggest a number of researchable questions. The first of these problems involves the identification of criteria for evaluating subsystem interchange. Variables affecting transaction between subsystems need to be identified. Klir proposed that attention be given to matters of system design rather than focusing on isolated health problems. Specifically, models could be developed by various health specialists that could be tested so that criteria might be established for a general system of health care. Health care organizations might be studied from the standpoint of how they tailor the patient care process to expressed community needs. Hearn raised some interesting questions about the study of ways that people emerging from the health care system in a relatively dependent state can be "recycled" into the growth-producing processes of society. Variables affecting that process need identification. These variables would relate to the great concern about quality of care and outcomes of care or services provided.

The questions are endless. The health research field is a viable one; applications of systems approaches have the promise of providing many new insights into extremely complex problems related to the health fields.

Index